Studies in Logic
Logic and Argumentation
Volume 68

Argument Technologies
Theory, Analysis, and Applications

Volume 58
Handbook of Mathematical Fuzzy Logic, Volume 3
Petr Cintula, Petr Hajek and Carles Noguera, eds.

Volume 59
The Psychology of Argument. Cognitive Approaches to Argumentation and Persuasion
Fabio Paglieri, Laura Bonelli and Silvia Felletti, eds

Volume 60
Absract Algebraic Logic. An Introductory Textbook
Josep Maria Font

Volume 61
Philosophical Applications of Modal Logic
Lloyd Humberstone

Volume 62
Argumentation and Reasoned Action. Proceedings of the 1st European Conference on Argumentation, Lisbon 2015. Volume I
Dima Mohammed and Marcin Lewiński, eds

Volume 63
Argumentation and Reasoned Action. Proceedings of the 1st European Conference on Argumentation, Lisbon 2015. Volume II
Dima Mohammed and Marcin Lewiński, eds

Volume 64
Logic of Questions in the Wild. Inferential Erotetic Logic in Information Seeking Dialogue Modelling
Paweł Łupkowski

Volume 65
Elementary Logic with Applications. A Procedural Perspective for Computer Scientists
D. M. Gabbay and O. T. Rodrigues

Volume 66
Logical Consequences. Theory and Applications: An Introduction.
Luis M. Augusto

Volume 67
Many-Valued Logics: A Mathematical and Computational Introduction
Luis M. Augusto

Volume 68
Argument Technologies: Theory, Analysis, and Applications
Floris Bex, Floriana Grasso, Nancy Green, Fabio Paglieri and Chris Reed, eds

Studies in Logic Series Editor
Dov Gabbay   dov.gabbay@kcl.ac.uk

Argument Technologies
Theory, Analysis, and Applications

Edited by
Floris Bex
Floriana Grasso
Nancy Green
Fabio Paglieri
Chris Reed

© Individual author and College Publications 2017
All rights reserved.

ISBN 978-1-84890-218-3

College Publications
Scientific Director: Dov Gabbay
Managing Director: Jane Spurr

http://www.collegepublications.co.uk

Printed by Lightning Source, Milton Keynes, UK

All rights reserved. No part of this publication may be reproduced, stored in a retrieval system or transmitted in any form, or by any means, electronic, mechanical, photocopying, recording or otherwise without prior permission, in writing, from the publisher.

Contents

Introduction. i

Part I: Theoretical Models

1. Defining the structure of arguments with an AI model of argumentation
Bin Wei & Henry Prakken. 1

2. A theoretical framework to support argumentation on information sources
Cristiano Castelfranchi, Rino Falcone, & Alessandro Sapienza 23

3. On the rationality of argumentative decisions
Fabio Paglieri. 39

Part II: Arguing on the Web

4. Arguers and the argument Web
Mark Snaith, Rolando Medellin, John Lawrence, & Chris Reed 57

5. From argumentation theory to social web applications
Lucas Carstens, Valentinos Evripidou, & Francesca Toni. 73

6. How e-minorities can argue on line. The case of Teatro Valle Occupato
Francesca D'Errico. 95

7. Positive and negative arguments in review systems: An approach with arguments
Simone Gabbriellini & Francesco Santini. 117

8. Hypotheses of analysis on the stylistic of arguments: A case study from Trip Advisor
Laura Bonelli. 131

Part III: Tools and Applications

9. ProtOCL: Specifying dialogue games using UML and OCL
Tangming Yuan & Simon Wells.. 151

10. Arguing about open source licensing issues on the web
Thomas F. Gordon... 165

11. Dialogues in US Supreme Court oral hearings
Latifa Al-Abdulkarim, Katie Atkinson, & Trevor Bench-Capon............. 193

12. An experience in documenting medical discussions through natural argumentation
Daniela Fogli, Massimiliano Giacomin, Fabio Stocco, & Federica Vivenzi..... 207

List of Contributors

Latifa Al-Abdulkarim
Department of Computer Science, University of Liverpool, UK
latifak@liverpool.ac.uk

Katie Atkinson
Department of Computer Science, University of Liverpool, UK
katie@liverpool.ac.uk

Trevor Bench-Capon
Department of Computer Science, University of Liverpool, UK
tbc@liverpool.ac.uk

Laura Bonelli
Istituto di Scienze e Tecnologie della Cognizione, Consiglio Nazionale delle Ricerche, Roma, Italy; "La Sapienza" University of Rome
laura.bonelli@istc.cnr.it

Lucas Carstens
Department of Computing, Imperial College London, UK
lucas.carstens10@imperial.ac.uk

Cristiano Castelfranchi
Institute for Cognitive Sciences and Technologies, ISTC-CNR, Rome, Italy
cristiano.castelfranchi@istc.cnr.it

Francesca D'Errico
Uninettuno University, Psychology Faculty, Rome, Italy
f.derrico@uninettunouniversity.net

Valentinos Evripidou
Department of Computing, Imperial College London, UK
evripidou10@imperial.ac.uk

Rino Falcone
Institute for Cognitive Sciences and Technologies, ISTC-CNR, Rome, Italy
rino.falcone@istc.cnr.it

Daniela Fogli
Department of Information Engineering, University of Brescia, Brescia, Italy
daniela.fogli@unibs.it

Simone Gabbriellini
Dipartimento di Economia e Management, Università di Brescia, Italy
simone.gabbriellini@unibs.it

Massimiliano Giacomin
Department of Information Engineering, University of Brescia, Brescia, Italy
massimiliano.giacomin@unibs.it

Thomas F. Gordon
Fraunhofer FOKUS, Berlin, Germany
thomas.gordon@fokus.fraunhofer.de

John Lawrence
School of Computing, University of Dundee, Dundee, DD1 4HN, UK
johnlawrence@computing.dundee.ac.uk

Rolando Medellin
School of Computing, University of Dundee, Dundee, DD1 4HN, UK
rmedellin@computing.dundee.ac.uk

Fabio Paglieri
Institute for Cognitive Sciences and Technologies, ISTC-CNR, Rome, Italy
fabio.paglieri@istc.cnr.it

Henry Prakken
Department of Information and Computing Sciences, Utrecht University and Faculty of Law, University of Groningen, The Netherlands
h.prakken@uu.nl

Chris Reed
Centre for Argument Technology, SSE, University of Dundee, Dundee DD1 4HN, Scotland, UK
c.a.reed@dundee.ac.uk

Francesco Santini
Dipartimento di Matematica e Informatica, Università di Perugia, Italy
francesco.santini@dmi.unipg.it

Alessandro Sapienza
Institute for Cognitive Sciences and Technologies, ISTC-CNR, Rome, Italy
alessandro.sapienza@istc.cnr.it

Mark Snaith
School of Computing, University of Dundee, DUNDEE, DD1 4HN, UK
marksnaith@computing.dundee.ac.uk

Fabio Stocco
Department of Information Engineering, University of Brescia, Brescia, Italy
fabio.stocco85@gmail.com

Francesca Toni
Department of Computing, Imperial College London, UK
ft@imperial.ac.uk

Federica Vivenzi
Department of Information Engineering, University of Brescia, Brescia, Italy;
federica.vivenzi@gmail.com

Bin Wei
School of Administrative Law, Southwest University of Political Science and Law, Chongqing, China
srsysj@126.com

Simon Wells
Edinburgh Napier University, Edinburgh, United Kingdom
s.wells@napier.ac.uk

Tangming Yuan
University of York, York, United Kingdom
tommy.yuan@york.ac.uk

Introduction

Floris Bex[1], Floriana Grasso[2], Nancy Green[3], Fabio Paglieri[4] & Chris Reed[5]

[1] Department of Information and Computing Sciences, Utrecht University, PO Box 80.089, 3508 TB, Utrecht, The Netherlands, f.j.bex@uu.nl
[2] Department of Computer Science, University of Liverpool, Ashton Building, Ashton Street, Liverpool L69 3BX,
room G.16, floriana@liverpool.ac.uk
[3] Department of Computer Science, University of North Carolina Greensboro, Greensboro, NC, USA 27402, nlgreen@uncg.edu
[4] Goal-Oriented Agents Lab (GOAL), Istituto di Scienze e Tecnologie della Cognizione, Consiglio Nazionale delle Ricerche (ISTC-CNR), Via San Martino della Battaglia 44, 00185 Roma, Italy, fabio.paglieri@istc.cnr.it
[5] Centre for Argument Technology, SSE, University of Dundee, Dundee DD1 4HN, Scotland, UK, c.a.reed@dundee.ac.uk

Over the last few decades, argument technologies has risen from being a highly technical sub-field of Artificial Intelligence to becoming a broad interdisciplinary research priority, with its own variety of sub-fields and methodologies. Initially, the philosophical tradition on argumentation was mostly appreciated in computer science as an effective technical solution to handle some key challenges in AI, such as the modelling of non-monotonic reasoning and the design of robust coordination protocols among large numbers of autonomous agents. Seminal contributions in this vein included, but were not limited to, Dung's abstract argumentation (1995), Pollock's work on defeaters (1995), and the range of proposals for multi-agent argumentation systems (for a comprehensive survey, see Rahwan & Simari, 2009). Soon these theoretical breakthroughs led to the development of interesting applications, and argument technologies started to attract interest as potentially effective tools for a variety of purposes. For instance, it became relevant to assess how the newly developed argumentation technologies might impact education (Andriessen, Baker, & Suthers, 2003), which naturally complemented the long-standing debate on how to promote critical thinking education (whose main current outcomes are included in Davies & Barnett, 2015) and the value of argumentation for learning, especially with respect to scientific inquiry (Driver, Newton, & Osborne, 2000).

Not surprisingly, such a variety of efforts and approaches led to the blossoming of multiple series of scholarly workshops and seminars (e.g., CMNA, Computational Models of Natural Argument, http://www.cmna.info/; ArgMAS, Argumentation in

Floris Bex, Floriana Grasso, Nancy Green, Fabio Paglieri, & Chris Reed

Multi-Agent Systems, http://www.mit.edu/~irahwan/argmas/), which since 2006 coalesced in the biennial international conference on Computational Models of Argument, COMMA (http://comma.csc.liv.ac.uk/), even though most of those workshops are still very much alive and well. More recently, all of these interconnected lines of research have shifted their focus towards a specific array of technologies – namely, online technologies. Given the impressive volume and dubious quality of online dialogical interactions people experience every day, it is no surprise that understanding how online technologies impact our social life has become a pressing concern. This is certainly true for people looking mostly at the benefits of such transformations, e.g., how it might allow better harnessing Surowiecki's alleged "wisdom of crowds" (2004); but the topic is equally urgent, if not more, for those who are worried by the shortcomings and bottlenecks online argumentation is likely to exhibit – e.g., the dubious value of Facebook as a learning platform (Kirschner, 2015) and the current outcry against the spreading of false information online (Del Vicario et al., 2016), typically attributed to the filtering effect of being selectively exposed to the views of like-minded peers (so called "echo chambers", see Sunstein, 2001).

Regardless of whether one is optimistic or pessimistic on the current status quo, the impact of online technology on the quality of social interaction poses key challenges that we need to collectively address: as some commentators recently put it, "too many mechanisms for online interaction hamper and discourage debate, facilitating poor-quality argument and fuzzy thinking. Needed are new tools, systems, and standards engineered into the heart of the Web to encourage debate, facilitate good argument, and promote a new online critical literacy" (Bex et al., 2016, p. 66). Such need is also attested by the booming interest in argument mining, that is, the use of Machine Learning and Natural Language Processing techniques to automatically analyze, categorize, and retrieve argument structures within online texts (e.g., next September the fourth edition of the successful ArgMining workshop will be held in Copenhagen, showcasing new advances in this field; https://argmining2017.wordpress.com/).

This volume takes stock of these developments in argument technologies, to offer a comprehensive and updated overview of some recent ramifications. The chapters are organized around three broad areas: theoretical models, arguing on the web, and tools and applications.

The contributions included in Part I, Theoretical Models, explore from different angles important aspects of argumentation theory, in relation to its application to argument technologies. Bin Wei and Henry Prakken focus on "Defining the structure of arguments with an AI model of argumentation": the structure of arguments is clearly a central issue in the field of informal logic and argumentation theory, and this paper discusses how the "standard approach" of Thomas, Walton, Freeman and others can be analyzed from a formal perspective. Wei and Prakken use the ASPIC+ framework for structured argumentation (Modgil & Prakken, 2014) for making the standard model of argument structure complete and for introducing a distinction between types of individual arguments and types of argument structures. Then they show that Vorobej's extension of the standard model with a new type of hybrid arguments (Vorobej,1995) is not needed if their formal approach is adopted, and they conclude by discussing the structure of so-called accrual of arguments.

In "A theoretical framework to support argumentation on information sources", Cristiano Castelfranchi, Rino Falcone and Alessandro Sapienza propose a theoretical connection between trust in information sources and argumentation on such sources. Given many sources of information asserting different things, deciding to accept or not a belief is a complicated task. Looking at the situation from the source's point of view, it is interesting to understand what are the variables influencing the final outcome, in order to manipulate them strategically: for a certain source, it could be easier to attack or support another source rather than just report an information. Leveraging the socio-cognitive theory of trust (Castelfranchi & Falcone, 2010), the authors describe and implement a theoretical and computational model that identifies all the variables playing a role when an agent has to evaluate whether to accept or not a given belief. Once identified these factors, all the arguments to support or to deny the belief can be traced back to them: this allows Castelfranchi and colleagues to present a principled list of possible argumentative maneuvers, to exemplify how the framework works.

The next paper is by Fabio Paglieri and it focuses "On the rationality of argumentative decisions": Paglieri first summarizes the basic assumptions of a decision-theoretic approach to argumentation, as well as some preliminary empirical findings based on that view, then he discusses the current relative neglect for decision making in argumentation theory. Much of this neglect seems to be based on the assumption that looking at arguments from a decision-theoretic perspective would lead to a merely descriptive approach: Paglieri argues against this claim, trying to show that, on the contrary, considering arguments as the product of decisions brings into play various normative models of rational choice. This presents argumentation theory and argument technologies with a novel challenge: how to reconcile strategic rationality with other normative constraints, such as inferential validity and dialectical appropriateness? It is suggested that strategic considerations should be included, rather than excluded, from the evaluation of argument quality, and this position is put in contact with the growing interest for virtue theory in argumentation studies (Aberdein & Cohen, 2016), as well as discussing its implications for designing and developing argument technologies.

Part II of the volume, Arguing on the Web, includes five contributions on various aspects of online argumentation, which brings multiple methodologies to bear on the problem of analyzing and supporting argumentative activity on the Internet. In "Arguers and the Argument", Mark Snaith, Rolando Medellin, John Lawrence and Chris Reed lay some groundwork in understanding the ways in which users of the Argument Web can be represented in the Argument Interchange Format (AIF, see Chesnevar et al., 2007). In particular, they tackle two specific challenges of (i) how to handle various roles on the Argument Web, based on the level and type of contribution they make; and (ii) how to support mixed initiative argumentation in which software agents may adopt a role defined by the previous activity of human arguers. This allows Snaith and colleagues to present an extra layer of the Argument Web, the Social Layer, which handles both the representation of users and connections to the social World Wide Web.

The same need to bridge the gap between argumentation theory and the social web is discussed by Lucas Carstens, Valentinos Evripidou and Francesca Toni in their contribution, "From argumentation theory to social web applications". The authors present their experience with developing concrete applications that make use of argumentation to various ends, including developments in social web debating settings, as well as potential uses of Natural Language Processing to support the proliferation of argumentative analysis. Based on this body of work, Carstens and colleagues discuss five challenges they consider crucial to the development of useful and usable argumentation applications. In particular, they look at a debating platform, *Quaestio-it.com* (Evripidou & Toni, 2014), and two decision-support systems, *Desmold* (http://www.desmold.eu/; see Baroni et al., 2015) and *Arg&Dec* (www.arganddec.com; see Cabanillas et al., 2013), to discuss whether and how they address these challenges and, if not, how they may do so in the future.

In the next paper, Francesca D'Errico discusses "How e-minorities can argue on line. The case of Teatro Valle Occupato". Her contribution aims at understanding argumentative moves in a minority discussing on line within the context of media-activism. The analysis is focused on the case study of some actors of the "Teatro Valle Occupato" (Occupied Valle Theater) in Rome, to show how activists use social media as a complementary tool to their actual protest, thus exploiting "e-tactics". Based on a previous content analysis of verbal change in social media (D'Errico, 2016), the present work analyses occupants and followers' argumentations in order to identify the quality of discussions and its possible empowerment effect. Results suggests that this e-minority is strongly focused on a "normative" and "cultural" core of arguments, with some differences in personal evaluations for what concerns activists. Implications in terms of interpersonal empowerment are also considered in the discussion.

Argumentation on social media is also the focus of the paper by Simone Gabriellini and Francesco Santini on "Positive and negative arguments in review systems: An approach with arguments". They study arguments in Amazon.com reviews, first by manually extracting positive (in favor of purchase) and negative (against it) arguments from each review concerning a selected product, and then linking such arguments to the rating score and length of reviews. For instance, they find that negative arguments are quite sparse during the first steps of the social review-process, while positive arguments are more equally distributed. Finally, they connect arguments through attack relations and compute Dung's (1995) extensions to check whether they capture such evolution through time.

Laura Bonelli also focuses on online reviews, in the paper "Hypotheses of analysis on the stylistic of arguments: A case study from Trip Advisor". She emphasizes how travelers' reviews on TripAdvisor and similar portals are based on persuasive arguments aimed at convincing other travelers of the truthfulness of the reviewers' experience and of the validity of their opinions. Starting from the idea that semantic and expressive levels of argumentation should not be seen as separate level of analysis, this chapter proposes an analytic framework of online sentiment analysis, which mainly focuses on the users' stylistic and rhetoric choices in a corpus of 100 Trip Advisor reviews of several hotels based in London, UK. The users' communicative behavior is analyzed using the taxonomy of emotive devices developed by Caffi and Janney (1994). Results show that most devices work as reinforcements of the users' inferred evaluative standpoints. The proposed analytic framework shows promise as a complement, rather than a competitor, to more traditional methods of argument analysis.

Part III, Tools and Applications, includes the last four contribution of the volume: the first two focus on important tools in making argument technologies work, whereas the other two papers present concrete applications of such technologies. Tangming Yuan and Simon Wells, in "ProtOCL: Specifying dialogue games using UML and OCL", notice how dialogue games (Hamblin, 1970; Walton & Krabbe, 1995) are becoming increasingly popular tools for Human-Computer Dialogue and Agent Communication. However, in spite of growing evidence of the value and utility of dialogue games, and also a range of novel implementations within specific problem domains, there remain very few tools to support the deployment of dialogue games based solutions within new problem domains. This paper introduces a new approach, called ProtOCL, to the specification of dialogue games, which adopts Unified Modeling Language (UML) and the Object Constraint Language (OCL) and enables the rapid movement from specification to deployment and execution. The dialogue game, DE, is used as an exemplar and is described using OCL to yield DE-OCL. Code generation is subsequently used to move from the DE-OCL description to executable code. This approach goes beyond existing description languages and their supporting tools by (1) using a description language that is familiar to a far larger user group, and (2) enabling code-generation using languages and technologies that are current industry standards.

In his paper "Arguing about open source licensing issues on the web", Thomas Gordon describes an interactive web application, called the MARKOS License Analyzer, developed in the European MARKOS project to help software developers to efficiently and cost effectively assess open source licensing issues and minimize their legal risks. The license analyzer is based on the Carneades Argumentation System (Gordon, Prakken, & Walton, 2007) and provides support for constructing, visualizing, evaluating and comparing competing legal arguments and theories. Arguments can be constructed both manually, via the Web interface, and automatically, from a knowledge-base consisting of an OWL ontology, a rulebase of argumentation schemes about open source software licensing issues, and facts in an RDF triplestore mined automatically from open source repositories such as GitHub.

The next paper also focuses on legal applications of argumentation tools, but from a very different perspective. In "Dialogues in US Supreme Court oral hearings", Latifa Al-Abdulkarim, Katie Atkinson and Trevor Bench-Capon argue that dialogue protocols in AI and law have become increasingly stylized, intended to examine the logic of particular legal phenomena such as burden of proof, rather than the procedures within which these phenomena occur (see also Bench-Capon, 1997). While such work has provided some valuable insights, this paper returns to the original idea of using dialogue moves to model particular procedures by examining some very particular dialogues: those found in oral hearings of the US Supreme Court. Al-Abdulkarim and colleagues characterize these dialogues, and illustrate the paper with examples taken from a close analysis of a case often modelled in AI and law, California v Carney, 1985. Thus this paper presents the preliminary investigation required to identify tools to provide computational support for the analysis of oral hearings.

The last contribution to this volume is by Daniela Fogli, Massimiliano Giacomin, Fabio Stocco and Federica Vivenzi, and it describes "An experience in documenting medical discussions through natural argumentation". The paper presents a prototypical tool to support the documentation of medical discussions, building on past experience in developing clinical decision support systems (Fogli & Guida, 2013). The authors detail how development of this tool came after a long phase of requirement specification and prototyping activity, which allowed identifying two main functionalities for the tool, corresponding to two components. A first component is devoted to support the user in documenting clinical discussions, exploiting a graphical notation tailored to physicians' habits. A second component exploits argumentation schemes in order to analyze documented discussions, possibly bringing to light weaknesses in the reasoning process.

Acknowledgements

Many people were instrumental in putting together this volume: some essays (chapters 2, 5, 6, 7, 8, and 10) were selected from a pool of 24 papers presented at the workshop on "Arguing on the Web 2.0", held in Amsterdam on June 30 and July 1, 2014, and later complemented with a second call for papers; other contributions (chapters 1, 3, 4, 9, 11, and 12) came from past editions of the Computational Models of Natural Argument workshop series (CMNA, http://www.cmna.info/). Regarding the Amsterdam event, it was chaired by Fabio Paglieri and Chris Reed, with Ulle Endriss acting as local organizer: we are very grateful to the University of Amsterdam for hosting the workshop, to the International Society for the Study of Argumentation (ISSA) for allowing it to be scheduled back-to-back with the ISSA 2014 conference, and to all the sponsors for their essential support – the European Network for Social Intelligence (SINTELNET, https://ec.europa.eu/digital-single-market/en/news/sintelnet-european-network-social-intelligence), the Center for Argumentative Technology of the University of Dundee, UK (ARG-tech, http://www.arg-tech.org/), and the Institute of Cognitive Sciences and Technologies of the National Research Council, Rome (ISTC-CNR, http://istc.cnr.it/). Most of the papers presented in Amsterdam that could not be included in this collection, due to length limitations, have appeared in a special issue of the journal *Philosophy &*

Technology, dedicated to "Online arguments: Theoretical and technological perspectives", and guest edited by Fabio Paglieri and Chris Reed (the introduction is available here: https://link.springer.com/article/10.1007/s13347-017-0264-4). As for the CMNA papers, the contribution by Wei and Prakken (chapter 1 of this volume) was originally presented in Montpellier, France, on August 27, 2012 at CMNA XII (http://cmna.csc.liv.ac.uk/workshops-12.html), which was co-located with the European Conference on Artificial Intelligence (ECAI); all other CMNA papers were instead presented in Rome, Italy, on June 13, 2013 at CMNA XIII (http://cmna.csc.liv.ac.uk/workshops-13.html), which took place as a joint workshop of the International Conference on AI in Law (ICAIL) and the Conference on User Modeling, Adaptation, and Personalization (UMAP). Last but not least, the present volume would have never been possible without the generous efforts of all authors and reviewers, as well as the unwavering support of the editorial staff of College Publications – most notably, Jane Spurr and John Woods. Even more essential was the careful editorial work generously performed by Silvia Felletti and Laura Bonelli, who turned a bunch of unedited drafts into the nicely formatted chapters you are about to peruse. To all these people we offer our sincere gratitude, in the hope that the contributions presented here will do justice to their expectations, by providing a thought-provoking bird-eye view of the current achievements and future perspectives of argument technologies.

References

Aberdein, A., & Cohen, D. (2016). Introduction: Virtues and arguments. *Topoi, 35(2)*, pp 339–343.

Andriessen, J., Baker, M., & Suthers, D. (Eds.) (2003). *Arguing to Learn: Confronting Cognitions in Computer-Supported Collaborative Learning Environments*. Dordrecht: Springer.

Baroni, P., Romano, M., Toni, F., Aurisicchio, M., & Bertanza, G. (2015). Automatic evaluation of design alternatives with quantitative argumentation. *Argument & Computation, 6(1)*, 24–49.

Bench-Capon, T. (1997). Argument in Artificial Intelligence and law. *Artificial Intelligence and Law, 5(4)*, 249–261.

Bex, F., Lawrence, J., Snaith, M., & Reed, C. (2013). Implementing the Argument Web. *Communications of the ACM, 56(10)*, 66–73.

Cabanillas, D., Bonada, F., Ventura, R., Toni, F., Evripidou, V., Carstens, L., & Rebolledo, L. (2013). A combination of knowledge and argumentation based system for supporting injection mould design. In K. Gibert, V. Botti, & R. Reig-Bolaño (Eds.), *Artificial Intelligence Research and Development. Proceedings of CCIA 2013* (pp. 293-296). Amsterdam: IOS Press.

Caffi, C., & Janney, R. (1994). Toward a pragmatics of emotive communication. *Journal of Pragmatics, 22*, 325–373.

Castelfranchi, C., & Falcone, R. (2010). *Trust Theory: A Socio-Cognitive and Computational Model*. Chichester: John Wiley & Sons.

Chesnevar, C., McGinnis, J., Modgil, S., Rahwan, I., Reed, C., Simari, C., South, M., Vreeswijk, G., & Willmott, S. (2007). Towards an argument interchange format. *The Knowledge Engineering Review, 21(4)*, 293–316.

Davies, M., & Barnett, R. (Eds.) (2015). *The Palgrave Handbook of Critical Thinking in Higher Education*. Berlin: Springer.

Del Vicario, M., Bessi, A., Zollo, F., Petroni, F., Scala, A., Caldarelli, G., Stanley, E., & Quattrociocchi, W. (2016). The spreading of misinformation online. *Proceedings of the National Academy of Sciences, 113(3)*, 554–559.

Driver, R., Newton, P., & Osborne, J. (2000). Establishing the norms of scientific argumentation in classrooms. *Science Education, 4(3)*, 287–312.

Dung, P. (1995). On the acceptability of arguments and its fundamental role in nonmonotonic reasoning, logic programming, and n-person games. *Artificial Intelligence, 77*, 321–357.

Evripidou, V., & Toni, F. (2014). Quaestio-it.com – a social intelligent debating platform. *Journal of Decision Systems, 23(3)*, 333–349.

Fogli, D., & Guida, G. (2013). Knowledge-centered design of decision support systems for emergency management. *Decision Support Systems, 55*, 336–347.

Gordon, T., Prakken, H., & Walton, D. (2007). The Carneades model of argument and burden of proof. *Artificial Intelligence, 171(10-15)*, 875–896.

Hamblin, C. (1970). *Fallacies*. London: Methuen.

Kirschner, P. (2015). Facebook as learning platform: Argumentation superhighway or dead-end street? *Computers in Human Behavior, 53*, 621–625.

Modgil, S., & Prakken, H. (2014). The ASPIC+ framework for structured argumentation: A tutorial. *Argument and Computation, 5(1)*, 31–62.

Pollock, J. (1995). *Cognitive Carpentry*. Cambridge: MIT Press.

Rahwan, I., & Simari, G. (Eds.) (2009). *Argumentation in Artificial Intelligence*. Berlin: Springer.

Sunstein, C. (2001). *Echo Chambers: Bush v. Gore, Impeachment, and Beyond*. Princeton/Oxford: Princeton University Press.

Surowiecki, J. (2004). *The Wisdom of Crowds*. New York: Doubleday.

Vorobej, M. (1995). Hybrid arguments. *Informal Logic, 17*, 289–296.

Walton, D., & Krabbe, E. (1995). *Commitment in Dialogue: Basic Concepts of Interpersonal Reasoning*. Albany: SUNY Press

Part I

Theoretical Models

Chapter 1

Defining the Structure of Arguments with an AI Model of Argumentation

Bin Wei[1] & Henry Prakken[2]

[1] School of Administrative Law, Southwest University of Political Science and Law, Chongqing, China, srsysj@126.com
[2] Department of Information and Computing Sciences, Utrecht University and Faculty of Law, University of Groningen, The Netherlands, h.prakken@uu.nl

Abstract. The structure of arguments is an important issue in the field of informal logic and argumentation theory. In this paper we discuss how the "standard approach" of Thomas, Walton, Freeman and others can be analyzed from a formal perspective. We use the *ASPIC+* framework for structured argumentation for making the standard model of argument structure complete and for introducing a distinction between types of individual arguments and types of argument structures. We then show that Vorobej's extension of the standard model with a new type of hybrid arguments is not needed if our formal approach is adopted. We finally discuss the structure of so-called accrual of arguments.

Introduction

The structure of arguments is an important issue in the field of informal logic and argumentation theory. Many logicians have given their definitions according to different criteria. The main issue is to define the different ways in which premises and conclusions can be combined to generate different structural argument types. The first model can be traced back to the works of Beardsley (1950), Thomas (1986) and Copi and Cohen (1990). Many informal logicians contributed to this topic, for instance, Walton (1996) and Freeman (2011). Vorobej (1995) extended their models with an additional argument type called "hybrid arguments". The main aim of this paper is to show how formal AI models of argumentation can be used to further extend and clarify these informal models of the structure of arguments. In particular, we argue that although these models provide much insight in the structure of argumentation, they still have some limitations, since their classifications are incomplete and since they do not distinguish between types of individual arguments and structures consisting of several arguments. Moreover, we argue that Vorobej's proposal can be clarified by making a distinction between deductive and defeasible arguments.

We aim to achieve our aims by applying the ASPIC+ framework of Modgil & Prakken (2011a,b), and Prakken (2010), since it arguably currently is the most general AI framework for structured argumentation. The framework has been shown (Gijzel & Prakken, 2011, Modgil & Prakken, 2011a,b; Prakken, 2010) to capture a number of other approaches to structured argumentation, such as assumption-based argumentation (Dung, Mancarella, & Toni, 2007), forms of classical argumentation (Gorogiannis & Hunter, 2011) and Carneades (Gordon, Prakken, & Walton, 2007). In Prakken (2010) it is also shown that *ASPIC+* can capture reasoning with presumptive argument schemes. The *ASPIC+* framework is based on two ideas: the first is that conflicts between arguments are often resolved with explicit preferences, and the second is that arguments are built with two kinds of inference rules: strict, or deductive rules, whose premises guarantee their conclusion, and defeasible rules, whose premises only create a presumption in favor of their conclusion. The second idea implies that *ASPIC+* does not primarily see argumentation as inconsistency handling in a given base logic: conflicts between arguments may not only arise from the inconsistency of a knowledge base but also from the defeasibility of the reasoning steps in an argument. Accordingly, arguments can in *ASPIC+* be attacked in three ways: on their uncertain premises or on their defeasible inferences, and the latter by either attacking their conclusion or the inference itself. We will use the *ASPIC+* framework to make four specific contributions: (1) to make the standard classifications complete; to (2) indicate and explain why convergent and divergent arguments are not arguments but argument structures; (3) to indicate and explain why Vorobej's class of hybrid arguments is not needed if an explicit distinction is made between deductive and defeasible arguments; and (4) to analyze the structure of so-called accrual of arguments.

This paper is organized as follows. In section 2, we introduce the standard informal model of argument structure and Vorobej's (1995) extension with so-called hybrid arguments. In section 3, we present a simplified version of the *ASPIC+* framework. We then use this framework in section 4 to complete the standard model and to distinguish between types and structures of arguments. In section 5, we discuss Vorobej's notion of hybrid arguments and how it can be captured in *ASPIC+*. In section 6, we define the structure of a special kind of argument, namely accrual of arguments. Section 7 concludes the paper.

Approaches to Argument Structure

We first introduce the main approaches to argument structures, notably the approach by e.g. Walton (1996) and Freeman (2011), which we will call the standard approach and Vorobej's (1995) extension with so-called hybrid arguments.

1.1 Standard Approach

The standard approach to the structure of arguments was introduced by Stephen N. Thomas (1986). He divided the arguments into (1) *linked arguments*, which means that every premise is dependent on the others to support the conclusion, (2) *convergent arguments*, which means that premises support the conclusion individually, (3) *divergent arguments*, which means that one premise supports two or more conclusions, and (4) *serial arguments*, which means that one premise supports a conclusion which supports another conclusion.

Walton (1996) then further discussed the structure of arguments. We present the informal definitions of the concepts of structures of arguments according to his latest description (2006).

Definition 1. The types of arguments are informally defined as follows:

(1) An argument is a *single argument* iff it has only one premise to give a reason to support the conclusion.
(2) An argument is a *convergent argument* iff there is more than one premise and where each premise functions separately as a reason to support the conclusion.
(3) An argument is a *linked argument* iff the premises function together to give a reason to support the conclusion.
(4) An argument is a *serial argument* iff there is a sequence $\{A_1,\cdots,A_n\}$ such that one proposition A_i acts as the conclusion drawn from other proposition A_{i-1} as premise and it also functions as a premise from which a new proposition A_{i+1} as conclusion is drawn.
(5) An argument is a *divergent argument* iff there are two or more propositions inferred as separate conclusions from the same premise.
(6) An argument is a *complex argument* iff it combines at least two arguments of types (2), (3), (4) or (5).

In order to show the diagrams of argument types and structures, we first need to define an inference graph. An inference graph is a labeled, finite, directed graph, consisting of statement nodes and supporting links indicating connecting relationships between nodes. In the diagrams of inference graphs, nodes are displayed as dots while supporting links are indicated using ordinary arrowheads. Then example diagrams of the above argument types are shown in Figure 1 (for simplicity, the distinction between strict and defeasible supporting links will be left implicit).

Example 1. Walton gives the following examples of, respectively, a convergent, divergent and linked argument:

(1) (A) Tipping makes the party receiving the tip feel undignified; (B) Tipping leads to an underground, black-market economy; (C) Tipping is a bad practice.
(2) (A) Smoking has been proved to be very dangerous to health; (B) Commercial advertisements for cigarettes should be banned; (C) Warnings that smoking is dangerous should be printed on all cigarette packages.
(3) (A) Birds fly; (B) Tweety is a bird; (C) Tweety flies.

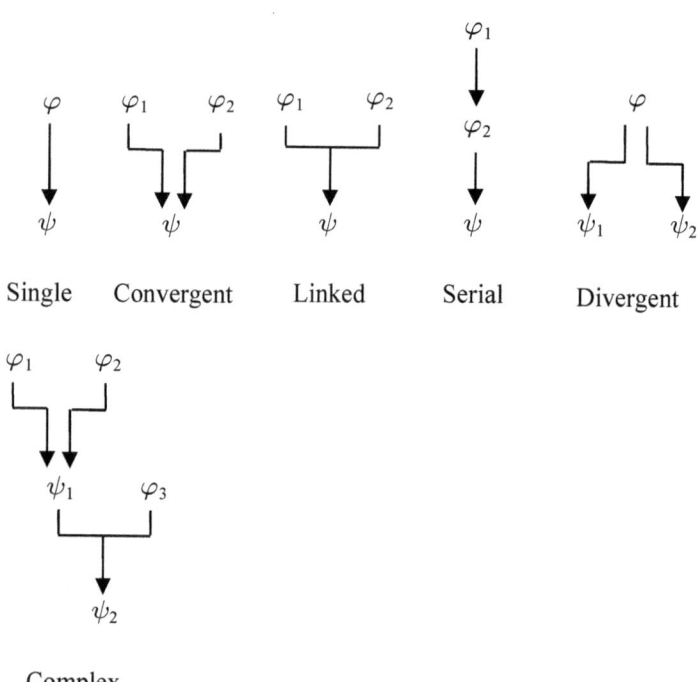

Figure 1. Diagrams of the Standard Approaches

In Example 1(1), the three statements form a convergent argument, since statements (A) and (B) function separately as a reason to support the conclusion (C). By contrast, in Example 1(2) these three statements form a divergent argument, since statement (B) and (C) are inferred as separate conclusions from the same premise (A). Finally, Example 1(3) is a linked argument, since neither premise alone gives any reason to accept the conclusion.

2.2. Hybrid Arguments

Mark Vorobej (1995) argued that the standard approach needs to be extended with a class of *hybrid* arguments. To discuss this class, we must first present Vorobej's basic definitions of types of arguments.

Definition2. An argument A is:
- *simple* iff A has exactly one conclusion. Otherwise, A is complex.
- *convergent* iff A is simple and each premise in A is relevant to C, where relevance is treated as a primitive dyadic relation obtaining in each instance between a set of propositions and a single proposition.

Definition3. A *linked* set and *linked* argument are defined as follows:
- A set of premises Δ forms a linked set *iff*
  1. Δ contains at least two members;
  2. Δ is relevant to C, and
  3. no proper subset of Δ is relevant to C.
- An argument A is *linked* iff A is simple and each premise in A is a member of some *linked* set.

Vorobej then motivates this new class of hybrid arguments with examples like the following one.

Example2. Consider example (F) as follows:
- (F): (1) All the ducks that I've seen on the pond are yellow. (2) I've seen all the ducks on the pond. (3) All the ducks on the pond are yellow.

Vorobej observes that (2) in isolation is not relevant to (3), so this is not a convergent argument. Secondly, (1) is relevant to (3), so (1) is not a member of any linked set, so this is also not a linked argument. Vorobej regards (F) as a hybrid argument, since (1) is relevant to the conclusion (3) and (2) is not relevant to the conclusion (3) but (1) and (2) together provide an additional reason for (3), besides the reason provided by (1) alone.

Vorobej provides the following definition of *hybrid* arguments in terms of a relation of supplementation between premises.

The relation of supplementation and *hybrid* argument are defined as follows:
- A set of premises Σ supplements a set of premises Δ iff
  1. Σ is not relevant to C;
  2. Σ is relevant to C;
  3. $\Sigma \cup \Delta$ offers an additional reason R in support of C, which Δ alone does not provide;
  4. Σ and Δ are the minimal sets yielding R which satisfy clauses (1),(2) and (3).

- An argument A is *hybrid* iff A is simple and contains at least one supplemented (or supplementing) set.

In Example2 premise (2) supplements premise (1). The argument is therefore a *hybrid* argument.

The *ASPIC+* Framework

The *ASPIC+* framework of Modgil & Prakken (2011a,b) and Prakken (2010) models arguments as inference trees constructed by two types of inference rules, namely, strict and defeasible inference rules. In this paper we use a simplified version of *ASPIC+* framework, with symmetric negation instead of an arbitrary contrariness function over the language and with just one instead of four types of premises. We also leave the various preference orderings on inference rules, the knowledge base and arguments implicit.

Definition5. [*Argumentation system*] An *argumentation system* is a tuple $AS = (\mathcal{L}, \mathcal{R})$, where

- is a logical language closed under negation ($\neg$). Below we write $\varphi = -\psi$ when either $\varphi = \neg\psi$ or $\psi = \neg\varphi$.
- $\mathcal{R} = \mathcal{R}_s \cup \mathcal{R}_d$ is a set of strict ($\mathcal{R}_s$) and defeasible ($\mathcal{R}_d$) inference rules such that $\mathcal{R}_s \cap \mathcal{R}_d = \emptyset$.

Definition6. [*Knowledge base*] A *knowledge base* in an argumentation system $(\mathcal{L}, \mathcal{R})$ is a set $\mathcal{K} \subseteq \mathcal{L}$.

Arguments can be constructed step-by-step by chaining inference rules into trees. In what follows, for a given argument the function *Prem* returns all its premises, *Conc* returns its conclusion *Sub* returns all its sub-arguments, while *TopRule* returns the last inference rule applied in the argument.

Definition7. [*Argument*] An argument A on the basis of a knowledge base $\mathcal{K}$ in an argumentation system $(\mathcal{L}, \mathcal{R})$ is:

1. φ if $\varphi \in \mathcal{K}$ with: $Prem(A) = \{\varphi\}$; $Conc(A) = \varphi$; $Sub(A) = \{\varphi\}$; $TopRule(A) = $ undefined.
2. $A_1, \cdots, A_n \to/\Rightarrow \psi$ if $A_1, \cdots, A_n$ are arguments such that there exists a strict/defeasible rule $Conc(A_1), \cdots, Conc(A_n) \to/\Rightarrow \psi$ in $\mathcal{R}_s \cup \mathcal{R}_d$.
   $Prem(A) = Prem(A_1) \cup, \cdots, \cup Prem(A_n)$;
   $Conc(A) = \psi$;
   $Sub(A) = Sub(A_1) \cup, \cdots, \cup Sub(A_n) \cup \{A\}$;
   $TopRule(A) = Conc(A_1), \cdots, Conc(A_n) \to/\Rightarrow \psi$
   $DefRules(A) = DefRules(A_1) \cup, \cdots, \cup DefRules(A_n)$.

An argument is *strict* if all its inference rules are strict and *defeasible* otherwise.

Definition 8. [*Maximal proper subargument*] Argument A is a *maximal proper subargument* of B iff A is a subargument of B and there does not exist any proper subargument C of B such that A is a proper subarugment of C.

The following example illustrates these definitions.

Example 3. Consider a knowledge base in an argumentation system with $\mathcal{R}_s = \{p, q \rightarrow s; u, v \rightarrow w\}$; $\mathcal{R}_d = \{p \Rightarrow t; s, r, t \Rightarrow v\}$; $\mathcal{K} = \{p, q, r, u\}$.

The diagram of the argument for w is displayed in Figure 2. Strict inferences are displayed with solid lines and defeasible inferences with dotted lines. Formally the argument and its subarguments are written as follows:

$A_1 = [p]$; $A_2 = [q]$; $A_3 = [r]$; $A_4 = [t]$; $A_5 = [m]$;
$A_6 = [A_1, A_2 \rightarrow s]$;
$A_7 = [A_3, A_4, A_6 \Rightarrow v]$;
$A_8 = [A_5 \rightarrow n]$;
$A_9 = [A_8 \Rightarrow u]$;
$A_{10} = [A_7, A_9 \rightarrow w]$.

We have that
$Prem(A_{10}) = \{p, q, r, t, m\}$;
$Conc(A_{10}) = w$;
$Sub(A_{10}) = \{A_1, A_2, A_3, A_4, A_5, A_6, A_7, A_8, A_9, A_{10}\}$;
$MaxSub(A_{10}) = \{A_7, A_9\}$;
$DefRules(A_{10}) = \{n \Rightarrow u; s, r, t \Rightarrow v\}$;
$Toprule(A_{10}) = \{u, v \rightarrow w\}$.

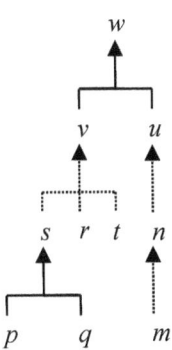

Figure 2. An Argument in *ASPIC+*

In the *ASPIC+* framework arguments can be attacked in three ways: attacking a premise, a defeasibly derived conclusion, or a defeasible inference. Attacks combined with an argument ordering yield a defeat relation, so that *ASPIC+* induces an abstract argumentation framework in the sense of Dung (1995). Since for present purposes the precise nature of the attack and defeat relations are irrelevant, we refer the reader for the formal definitions to Modgil & Prakken (2011) and Prakken (2010).

Types and Structures of Argument

We now give a new classification of arguments in terms of the *ASPIC+* framework and then define so-called argument structures, which are collections of arguments with certain features. We first define two kinds of *unit* arguments and then define several other argument notions consisting of these two *unit* types in different ways. We finally define various structures of argument in terms of the various definitions of argument types.

Definition 9. [*Argument type*] The types of arguments can be defined as follows:

(1) An argument A is an *unit I* argument iff A has the form $B \rightarrow \psi$ or $B \Rightarrow \psi$ and subargument B is an atomic argument $B:\varphi$. We call the inference rule $\varphi \rightarrow \psi$ or $\varphi \Rightarrow \psi$ an *unit I* inference.

(2) An argument A is an *unit II* argument iff A has the form $B_1, \cdots, B_n \rightarrow \psi$ or $B_1, \cdots, B_n \Rightarrow \psi$ and subarguments $B_1, \cdots, B_n$ are atomic arguments $B_1:\varphi_1, \cdots, B_n:\varphi_n$. We call the inference rule $\varphi_1, \cdots, \varphi_n \rightarrow \psi$ or $\varphi_1, \cdots, \varphi_n \Rightarrow \psi$ an *unit II* inference.

(3) An argument A is a *multiple unit I* argument iff all inferences $r_1, \cdots, r_n$ in the argument A are *unit I* inferences.

(4) An argument A is a *multiple unit II* argument iff all inferences $r_1, \cdots, r_n$ in the argument A are *unit II* inferences.

(5) An argument A is a *mixed* argument iff A has at least one *unit I* subargument and *unit II* subargument.

We display the diagrams of argument types in Figure 3. For simplicity, we assume $n=2$ in these diagrams and show only one case of a *mixed* argument.

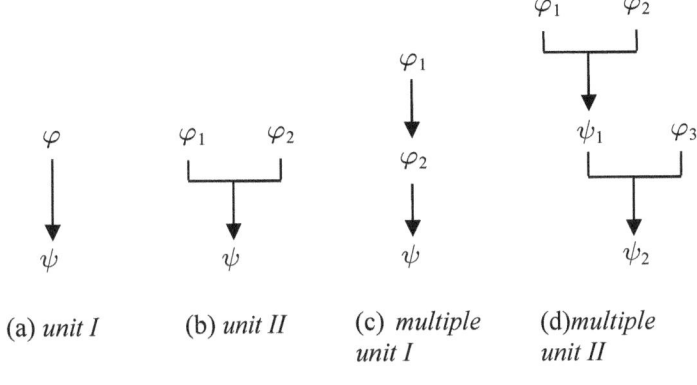

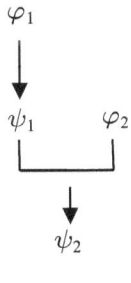

(d)*mixed*

Figure 3. Diagrams of Argument Types in the New Classification

Proposition1. Every argument is of exactly one argument type.

Proof. Firstly, we prove the existence of an argument type by induction on the number of unit inferences. For $n=1$, argument A corresponds to an *unit I* argument. For $n=k>1$, argument A corresponds to a *multiple unit I* argument, a *multiple unit II* argument, or a *mixed* argument. For $n=k+1$, we represent argument A as $B_1,\cdots,B_n \Rightarrow \psi$, where $m \leq n$. Consider the following cases:

(1) If B_i is a *multiple unit I* argument and r_{k+1} is an *unit I* inference, then according to definition7 and definition9(3), A is a *multiple unit I* argument.

(2) If B_i is a *multiple unit I* argument and r_{k+1} is an *multiple unit II* inference, then according to definition7 and definition9(5), A is a *mixed* argument.

(3) If B_i is a *multiple unit II* argument and r_{k+1} is an *unit I* inference, then according to definition7 and definition9(5), A is a *mixed* argument.
(4) If B_i is a *multiple unit II* argument and r_{k+1} is an *unit II* inference, then according to definition7 and definition9(4), A is a *multiple unit II* argument.
(5) If B_i is a *mixed* argument and r_{k+1} is an *unit I* or *unit II* inference, then according to definition7 and definition9(5), A is a *mixed* argument.

Secondly, we prove the property of uniqueness of argument type. Assume there exists an argument A corresponding to two or more argument types: then there must exist two or more top rules in the argument, and then there are two or more conclusions in A, which contradict the definition of argument.

Consider again Example 3. We have that A_1, A_2, A_3, A_4, A_5 are atomic arguments, A_8 is an *unit I* argument, A_6 is an *unit II* argument, A_9 is a *multiple unit I*, A_7 is a *multiple unit II*, and A_{10} is a *mixed* argument.

We next define several argument structures, which are sets of arguments with certain properties. We should first define connected arguments and an interconnected argument set as follows:

Definition10. Argument A and B are connected iff there exist $A' \in Sub(A)$ and $B' \in Sub(B)$, such that $Conc(A') \in Prem(B')$ or $Conc(A')=Conc(B')$ or $Prem(A') \subseteq Prem(B')$.

Proposition2. An argument A is connected with any of its subarguments.

Proof. For any $A_i \in Sub(A)$ it holds that $Prem(A_i) \subseteq Prem(A)$. From definition10, it follows that A_i is connected with A.

In Example 3, argument A_{10} is connected with any subargument A_i, where $i \in \{1,\cdots,9\}$. But minimal subarguments of A_{10} are not connected.

Definition11. A set of arguments $S=\{A_1,\cdots,A_n\}$ ($n \geq 2$) is *interconnected* iff for any argument A_i and $A_j \in S$, there exists a sequence of arguments $B_1,\cdots,B_k$, $B_{k+1},\cdots,B_m$ in S, where B_k is connected with B_{k+1} ($1 \leq m \leq n$), such that A_i is connected with B_1 and A_j is connected with B_m.

In Example3, $\{A_1,\cdots,A_{10}\}$ is interconnected. Moreover, let $S=\{A, B, A', B'\}$, where $A=[p]$, $B=[q]$, $A'=[A \Rightarrow p]$ with inference rule $p \Rightarrow q$ and $B'=[B \Rightarrow q]$ with inference rule $r \Rightarrow s$. From definition11, it follows that S is not interconnected, since (1) there is no argument to connect A with B or B' and (2) there is no argument to connect B with A' and (3) there is no argument to connect A' and B'.

Corollary1. A set of arguments S consisting of an argument A and all of its subarguments is interconnected.
Proof. Follows from proposition2 and definition11.

We call the argument set which consists of an argument and its subarguments as classic interconnected set and non-classic interconnected set otherwise. For instance, in Example3, $\{A_1,\cdots,A_{10}\}$ is classic interconnected.

Definition12. The set of argument structures[1] is defined as follows:

(1) A set of arguments $\{A_1,\cdots,A_n\}$ is a *serial convergent structure* SCS iff there are only *unit I* arguments in the set of arguments $\{A_1,\cdots,A_n\}$ and for any A_i and A_j we have $Conc(A_i)=Conc(A_j)$, where $i \neq j$.

(2) A set of arguments $\{A_1,\cdots,A_n\}$ is a *serial divergent structure* SDS iff there are only *unit I* arguments in the set of arguments $\{A_1,\cdots,A_n\}$ and for any A_i and A_j we have $Prem(A_i)=Prem(A_j)$, where $i \neq j$ and $A_i \not\in Sub(A_j)$.

(3) A set of arguments $\{A_1,\cdots,A_n\}$ is a *linked convergent structure* LCS iff it contains only *unit II* arguments and for any A_i and A_j we have $Conc(A_i)=Conc(A_j)$, where $i \neq j$.

(4) A set of arguments $\{A_1,\cdots,A_n\}$ is a *linked divergent structure* LDS iff it contains only *unit II* arguments and for any A_i and A_j we have $Prem(A_i)=Prem(A_j)$, where $i \neq j$.

(5) A set of arguments $\{A_1,\cdots,A_n\}$ is a *mixed structure MS* iff it is non-classic interconnected and it is not of the form of either SCS, SDS, LCS or LDS.

We display the diagrams of argument structures in Figure4. For simplicity, we assume $n=2$ in the diagrams and show only one case of *mixed structure*.

---

[1] The structure here is different from the structure in informal approaches, where it refers to the structure of an *individual* argument.

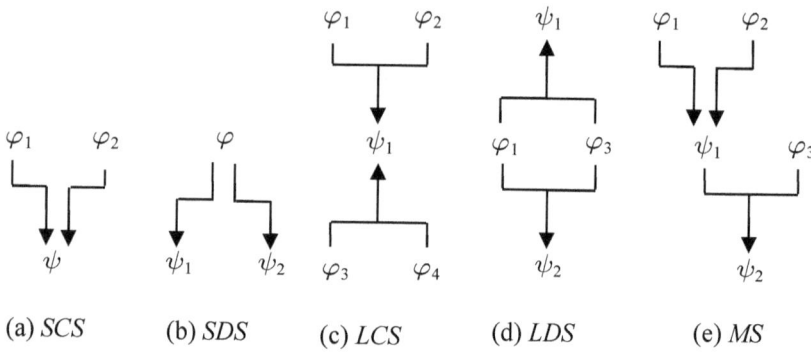

Figure 4. Diagrams of Argument structures in the New Classification

Proposition3. All of argument structures (*SCS*, *SDS*, *LCS*, *LDS* or *MS*) are non-classic interconnected.
Proof. Follows from definition 11 and definition 12.

Proposition4. If a set of arguments S is a *mixed* structure, then there exist at least one *unit I* argument and one *unit II* argument in it.
Proof. Suppose for contradiction that there does not exist *unit I* argument and one *unit II* argument in a set of arguments S. From definition 9, we have all cases as follows:

S does not contain *unit I* argument and one *unit II*. It contradicts definition 9, since all of argument types consists of *unit I* argument or one *unit II*.

S only contains *unit I* argument. If S is not interconnected, then S is not in any argument structure. If S is interconnected, then S is a *SCS* or *SDS* according to definition 12.

S only contains *unit II* argument. If S is not interconnected, then S is not in any argument structure. If S is interconnected, then S is a *LCS* or *LDS* according to definition 12.

Corollary2. If any two proper subsets of interconnected argument set S are of two different types of *SCS*, *SDS*, *LCS* and *LDS*, then S is a *mixed structure*.
Proof. Follows from proposition 4.

4.1. Reconsidering the Standard Approach

First, we consider the correspondence between the standard approach and our new approach. It is easy to see that single, linked and serial arguments, respectively, correspond to *unit I*, *unit II* and *multiple unit I* arguments.

However, *convergent* and *divergent* arguments are not arguments any more, since a *convergent* "argument" now is an argument structure consisting of a number of distinct *unit I* arguments for the same conclusion, while a divergent "argument" now is an argument structure consisting of a number of distinct *unit I* argument with the same premise. For instance, in Example1(1) there are two arguments $A \Rightarrow (C)$ and $B \Rightarrow (C)$ for the same conclusion (C), and in Example1(2), there are two arguments $A \Rightarrow (B)$ and $A \Rightarrow (C)$ with the same premise (A) where but different conclusions.

	Argument types					
Standard approach	single	linked	serial		complex
New approach	unit I	unit II	multiple unit I	multiple unit II	mixed

Table 1. Comparison of Argument Types

	Argument structures					
Standard approach	convergent	divergent			complex
New approach	serial convergent structure	serial divergent structure	linked convergent structure	linked divergent structure	mixed structure

Table 2. Comparison of Argument Structures

Therefore, the classes of convergent, divergent "arguments" are not arguments but argument *structures*. Actually, they correspond to the serial convergent structure *SCS* and the serial divergent structure *SDS*. Moreover, the class of complex arguments in the standard approach is not an argument if it contains *SCS* or *SDS*, but instead corresponds to the mixed argument structure *MS*. Otherwise, it corresponds to a mixed argument.

From the above analysis we see that the standard approach is incomplete and, moreover, does not distinguish types of individual argument from types of argument structures. We can show the comparisons between the standard approach and our new approach in *Table1* and *Table2*. We can also conclude that the new classification in terms of the *ASPIC+* framework is helpful in clarifying and complementing the standard approach.

5. The Problem of Hybrid Arguments

In this section we analyze why Vorobej's class of hybrid arguments is not needed if our approach is adopted. In our new approach, Vorobej's hybrid "argument" are not arguments but argument structures consisting a number of arguments. More specifically, they are of type mixed structure MS or linked convergent structure LCS.

We first make a notion explicit and redefine a definition. In Vorobej (1995) the notion of relevance is implicit and treated as a primitive dyadic relation. We note that there are two kinds of relevance: *defeasible relevance* indicates the support from a set of arguments to the conclusion via a defeasible inference, while *strict relevance* indicates the support form a set of arguments to the conclusion via a strict inference.

In the *ASPIC+* framework, we write $S \vdash \varphi$ if there exists a strict argument for φ with all premises taken from S, and $S|\sim\varphi$ if there exists a defeasible argument for φ with all premises taken from S. Then Definition 4 can be rewritten as follows:

Definition 13. A set of premises Σ supplements a set of premises Δ iff (1) $\Sigma \not\vdash C$ and $\Sigma \neq \emptyset$; (2) $\Delta|\sim C$; (3) $\Sigma \cup \Delta \vdash C$ or $\Sigma \cup \Delta|\sim C$, and (4) $\Sigma \cup \Delta$ is the minimal set satisfying clauses (1),(2) and (3) when $\cup \Delta \vdash C$.

If a set of premises $\Sigma = \{P_1, \ldots, P_m\}$ supplements a set of premises $\Delta = \{Q_1, \ldots, Q_n\}$, then we have two arguments A and B, where argument A is of the form $Q_1, \ldots, Q_n \Rightarrow C$ and argument B is of the form $P_1, \ldots, P_m, Q_1, \ldots, Q_n \Rightarrow C$ or $P_1, \ldots, P_m, Q_1, \ldots, Q_n \rightarrow C$.

Thus, the hybrid argument here is a (1) mixed structure *MS* consisting of a *unit I* argument and a *unit II* argument, if $m=1$, or (2) a linked convergent structure *LCS* consisting of two linked arguments, if $m>1$.

We now first reconsider Example 2.
- (F): (1) All the ducks that I've seen on the pond are yellow. (2) I've seen all the ducks on the pond. (3) All the ducks on the pond are yellow.

Arguably, (1) supports (3) because of the defeasible inference rule of enumerative induction:

- All observed F's are G's $\Rightarrow$ all F's are G's.

Moreover, (1) and (2) together arguably support (3) because of a deductive version of enumerative induction:
- All observed F's are G's, all observed F's are all F's $\Rightarrow$ all F's are G's.

We then see that the apparently hybrid argument is in fact a convergent structure consisting of two separate arguments for the same conclusion, sharing one premise:
- $A = [1 \Rightarrow (3)]$ with a defeasible inference rule: All observed F's are G's $\Rightarrow$ all F's are G's;
- $B = [1,2 \Rightarrow (3)]$ with a strict inference rule: All observed F's are G's, all observed F's are all F's $\rightarrow$ all F's are G's.

Actually, all examples in Prakken (2010) can be reconstructed in terms of these two kinds of structures:

Example4. Consider examples (G) and (J) as follows:

(G): (1) My duck is yellow. (2) Almost without exception, yellow ducks are migratory. (3) My duck is no exception to any rule. (4) My duck migrates.

(H): (1) My duck is yellow. (2) Most yellow ducks, especially those born in Ontario, are migratory. (3) My duck was born in Enterprise. (4) Enterprise is in Ontario. (5) My duck is migratory.

In example (G), we have that $\{(1),(2)\}|\sim(4)$ and $\{(1),(2),(3)\}\vdash(4)$, so we have two arguments A and B for the same conclusion:
- $A = [1,2 \Rightarrow (4)]$ with a defeasible inference rule: almost without exception X's are Y's, a is a $X \Rightarrow a$ is a Y;
- $B = [1,2,3 \rightarrow (4)]$ with a strict inference rule: almost without exception X's are Y's, a is a X, a is no exception to any rule $X \rightarrow a$ is a Y.

In example (H), there are two arguments A and B based on $\{(1),(2)\}|\sim(5)$ and $\{(1),(2),(3),(4)\}|\sim(5)$:
- $A = [1,2 \Rightarrow (5)]$ with a defeasible inference rule: Most X's are Y's, especially X's born in Z, a is a $X \Rightarrow a$ is a Y;
- $B = [1,2,3,4 \Rightarrow (5)]$ with a defeasible inference rule: Most X's are Y's, especially X's born in Z, a is a X, a born in y, y is in $Z \Rightarrow a$ is a Y.

On our account arguments in example (G) and (H) are both linked convergent structures.

Example5. Consider examples (H) and (I) as follows:
- (I): (1) All the ducks that Data has seen on the pond are yellow. (2) All the ducks that Dax has seen on the pond are yellow. (3) Data has seen 96% of the ducks on the pond. (4) All the ducks on the pond are yellow.

- (J): (1) Data quacks. (2) Data has webbed feet. (3) 95% of those creatures who both quack and have webbed feet are ducks. (4) Data is a duck.

In example (I), there are two arguments A, B based on $\{(1)\}|\sim(4)$, $\{(1),(3)\}|\sim(4)$ (Note that argument based on $\{(1),(2),(3)\}|\sim(4)$ is not an argument, since $\{(1),(2)\}$ is not the minimal set yielding (4) and then (3) does not supplement $\{(1),(2)\}$:
- $A = [1 \Rightarrow (4)]$ with a defeasible inference rule: All observed F's are G's $\Rightarrow$ all F's are G's;
- $C = [1, 3 \Rightarrow (4)]$ with a defeasible inference rule: All observed F's are G's, 95% observed $F \Rightarrow$ all F's are G's.

In example (J), there are four arguments A, B, C and D based on $\{(1)\}|\sim(4)$, $\{(2)\}|\sim(4)$, $\{(1),(2)\}|\sim(4)$ and $\{(1),(2),(3)\}|\sim(4)$:
- $A = [1 \Rightarrow (4)]$ with a defeasible inference rule: x quacks $\Rightarrow x$ is a duck;
- $B = [2 \Rightarrow (4)]$ with a defeasible inference rule: x has webbed feet $\Rightarrow x$ is a duck;
- $C = [1, 2 \Rightarrow (4)]$ with a defeasible inference rule that aggregates the two previous inference rules;
- $D = [1, 2, 3 \Rightarrow (4)]$ with a defeasible inference rule: a is a Y, a is a Z, 95% of x's who are both Y and Z are $T \Rightarrow a$ is a T.

On our account arguments in example (I) is a linked convergent structure and (J) is a mixed structure.

Example 6. Consider example (K) as follows:
- (K): (1) Data and Dax have the same diet. (2) Data and Dax receive the same amount of exercise. (3) Data is a healthy duck. (4) Dax is a healthy duck.

In example (K), there are three arguments A, B and C based on $\{(1),(3)\}|\sim(4)$, $\{(2),(3)\}|\sim(4)$ and $\{(1),(2),(3)\}|\sim(4)$ (Note that no matter $\{(2)\}$ supplements $\{(1),(3)\}$ or $\{(1)\}$ supplements $\{(2),(3)\}$, it would follow $\{(1),(2),(3)\}|\sim(4)$):
- $A = [1,3 \Rightarrow (4)]$ with a defeasible inference rule: x and y have the same diet, x is a healthy duck $\Rightarrow y$ is a healthy duck;
- $B = [2,3 \Rightarrow (4)]$ with a defeasible inference rule: x and y receive the same amount of exercise, x is a healthy duck $\Rightarrow y$ is a healthy duck;
- $C = [1,2,3 \Rightarrow (4)]$ with a defeasible inference rule that aggregates the two previous inference rules.

On our account arguments in example (K) is a linked convergent structure. The diagrams of the arguments in above examples are displayed in Figure 5.

Defining the structure of arguments with an AI model of argumentation

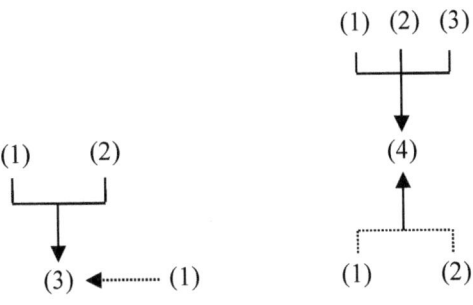

(a) *Example F*

(b) *Example G*

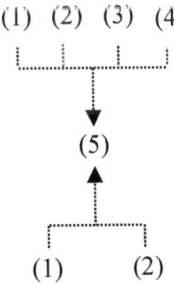

(c) *Example H*

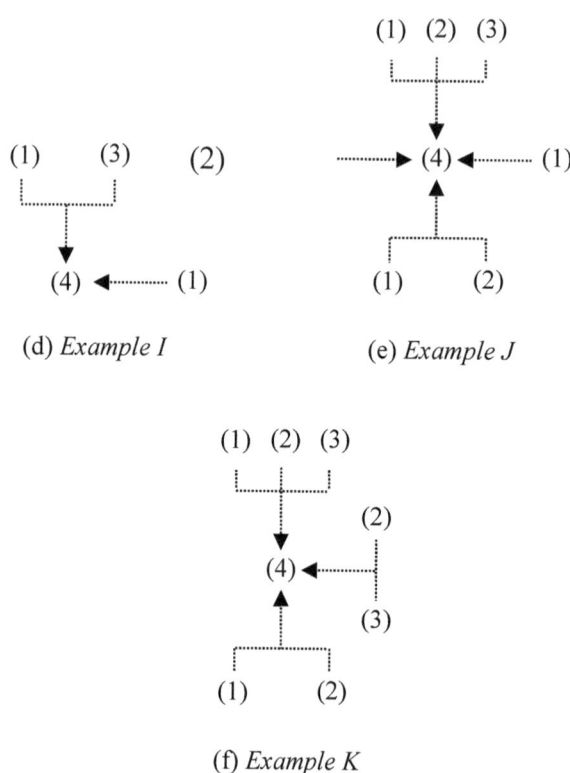

Figure 5. Diagrams of the Vorobej's examples

6. The Problem of Accrual of Arguments

In this section, we discuss the structure of so-called accrual of arguments. It often happens in natural arguments that several reasons support one proposition in such a way that each reason is offered as additional support for the proposition. This is called accrual of reasons or accrual of arguments.

In the AI literature many contributions (Gomez et al., 2009; Prakken, 2005; Verheij, 1995) focus on the formalisation of accrual of arguments. For example, Prakken (2010) presents the accrual of arguments as a form of inference in a standard logical framework for defeasible argumentation.

We first show an example of accural of arguments. Suppose that P is "witness a testifies φ" and Q is "witness b testifies φ", then we have two arguments:

$A=[A_1 \Rightarrow \varphi]$, where $A_1=[P]$ and $B=[B_1 \Rightarrow \varphi]$, where $B_1=[Q]$. Thus these two arguments have the same conclusion \varphi with defeasible rule r: $testifies(x,\varphi) \Rightarrow \varphi$. According to the treatment of Prakken (2005) accrual is a new form of inference, and the basic idea towards formalizing this special inference can be divided into two points.

First, the conclusion of each individual defeasible inference step is labelled with the premises of the applied defeasible inference rule. Given a set of defeasible inference rules, they are slightly reformulated to the effect that their conclusions are labelled with the set of their premises. So, defeasible modus ponens for $\Rightarrow$ can be defined as follows:

$$\varphi, \varphi \Rightarrow \psi | \sim \psi^{\{\varphi,\ldots,\varphi \Rightarrow \psi\}}.$$

Second, a new defeasible inference rule is introduced that takes any set of labelled versions of a certain formula and produces the unlabelled version:

$$\varphi^{l_1}, \ldots, \varphi^{l_n} | \sim \varphi$$

Note that the above labels will for readability often be abbreviated to $l_1, \ldots, l_n$ and the rule is a scheme for any natural number i such that $1 \leq i \leq n$. Therefore, in the above example we have a new inference rule: $\varphi^{P,P \Rightarrow \varphi}, \varphi^{Q,Q \Rightarrow \varphi} | \sim \varphi$ and the accrual of arguments of this example can be presented as $C=[A,B \Rightarrow \varphi]$, where $A=[A_1 \Rightarrow \varphi]$ and $B=[B_1 \Rightarrow \varphi]$.

The accrual of arguments as formalized in Modgil and Prakken (2011) is different from the notion of convergent arguments in informal approaches, since as we have discussed, the so called convergent argument is an argument structure rather than an argument. Actually, accrual of arguments is a mixed argument or *multiple unit II* argument with defeasible inference rule $\varphi^{l_1}, \ldots, \varphi^{l_n} | \sim \varphi$ as toprule. It should be noted that all subarguments of the accrual of arguments only have defeasible top rules, since accrual would not make sense for strictly derived conclusions. Then we can conclude the definition of the accrual of arguments as follows:

Definition 13. An argument A is an accrual of arguments with conclusion φ iff A is a *multiple unit II* argument or *mixed* argument with $Toprule(A)= \varphi^{l_1}, \ldots, \varphi^{l_n} \Rightarrow \varphi$.

We now reconsider example (K) and show the accrual of arguments inside. According to the analysis in Example 6, there are two linked arguments for a same conclusion with two inferences rules:
- r_1: x and y have the same diet, x is a healthy duck $\Rightarrow$ y is a healthy duck;

- r_2: x and y receive the same amount of exercise, x is a healthy duck
  $\Rightarrow y$ is a healthy duck.

Arguably, there is an accrual argument for statement (4), accruing two arguments A and B for (4) based on two labelled versions of (4), viz. $(4)^{\{(1),(3),r_1\}}$ and $(4)^{\{(2),(3),r_2\}}$. Applying the accrual rule: $(4)^{\{(1),(3),r_1\}}$, $(4)^{\{(2),(3),r_2\}}|\sim(4)$ to the above two reasons results in a defeasible argument for (4), namely accrual argument C, which can be represented as $C=[A,B\Rightarrow(4)]$, where $A=[1,3\Rightarrow(4)^{\{(1),(3)\}}]$ and $B=[2,3\Rightarrow(4)^{\{(2),(3)\}}]$. Actually, C is a *multiple unit II* argument and its structure is shown in Figure 6:

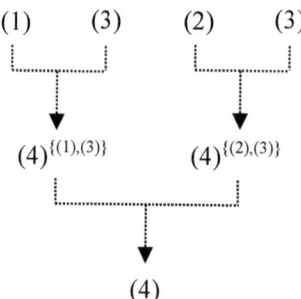

Figure 6. An accrual of arguments

Proposition5. If a non-classic interconnected argument set S contains an argument A which is an accrual of arguments, then
- it is a *mixed* structure, if S contains both *unit I* inferences and *unit II* inferences; or
- it is a *linked convergent* structure, if S only contains *unit II* inferences.

Proof. Firstly, since argument A is an accrual of arguments, then by definition 14, A contains *unit II* inferences.

Secondly, suppose for contradiction that S is not a mixed structure. By assumption and definition 12, S should be *SCS*, *LCS*, *SDS* or *LDS*, since S is non-classic interconnected. But S contains either *unit I* or *unit II*, contradicting S contains both *unit I* inferences and *unit II* inferences.

Thirdly, if S only contains *unit II* inferences and it is non-classic interconnected, then by definition 12 it is easy to conclude that S is a *linked convergent* structure.

Moreover, the interconnected argument set S consists of arguments in example (K) which contains three arguments in Example 6 and one accrual of arguments we have discussed. From the above proposition, since S only contains *unit II* inferences, we have that S is a *linked convergent* structure.

Corollary3. If an argument is an accrual of arguments, then it must belong to a set of arguments which is a *LCS* or *MS*.

Proof. Follows from proposition 5.

7. Conclusion

In this paper we showed how AI models of argumentation can be used to clarify and extend informal-logic approaches to the structure of arguments. We indicated that the standard approach is incomplete and then defined a complete classification of types of arguments in terms of the *ASPIC+* framework. We highlighted that *convergent* and *divergent* "arguments" in the standard approach are not arguments but sets of arguments, which we have classified as argument structures. We also showed that Vorobeij's *hybrid* arguments can be defined in terms of our classification if the distinction between deductive and defeasible inferences is made explicit, thus obviating the need to introduce a new type of argument to handle the examples discussed by Vorobeij. Finally, we applied the new approach to analyze the structure of accrual of arguments and we defined it as two possible argument types. We believe that our contributions are particularly relevant for argumentation theory, since we have clarified and extended terminology concerning classifications of arguments which is often used by argumentation theorists. Thus we have shown how formal methods can be of use not just in AI but also in argumentation theory and informal logic.

Acknowledgments

We thank the anonymous reviewers for their useful comments on the earlier versions of this paper. Bin Wei was supported by the Humanity and Social Science Youth foundation of Ministry of Education of China (No.15YJCZH182), the Education Scientific Planning Project of Chongqing (No.2015-GX-018) and by the Major Program of the National Social Scinence Fund (No.15AZX020).

References

Beardsley, M. (1950). *Practical Logic*. Englewood Cliffs: Prentice-Hall, 1950.

Bondarenko, A., Dung, P.M., Kowalski, R.A., & Toni. F. (1997). An abstract, argumentation-theoretic approach to default reasoning. *Artificial Intelligence*, 93, 63-101.

Copi I.M., & Cohen, C. (1990). *Introduction to Logic. 8th edition*. Macmillian Publishing Company, 18-22.

Dung, P.M. (1995). On the acceptability of arguments and its fundamental role in nonmonotonic reasoning, logic programming, and n-person games. *Artificial Intelligence*, 77: 321-357.

Dung, P.M., Mancarella, P., & Toni. F. (2007). Computing ideal sceptical argumentation. *Artificial Intelligence, 171*, 642-674.

Freeman, J.B. (2011). Argument Structure: Representation and Theory. Springer, 148.

Gijzel, B. van., & Prakken, H. (2011). Relating Carneades with abstract argumentation. *Proceedings of the 22nd International Joint Conference on Artificial Intelligence*, AAAI Press, 1113-1119.

Gijzel, B. van., & Prakken, H. (2012). Relating Carneades with abstract argumentation via the *ASPIC+* framework for structured argumentation. *Argument and Computation, 3(1)*, 21-47.

Gomez Lucero, M. J., Chesnevar, C. I., & Simari, G. R. (2009). On the accrual of arguments in Defeasible Logic Programming. *Proceedings of the 21st International Joint Conference on Artificial Intelligence*, AAAI Press, 804-809.

Gordon, T.F., Prakken, H. & Walton, D.N. (2007). The Carneades model of argument and burden of proof. *Artificial Intelligence, 171*, 875-896.

Gorogiannis, N., & Hunter, A. (2011). Instantiating abstract argumentation with classical-logic arguments: postulates and properties. *Artificial Intelligence, 175*, 479-1497.

Modgil. S. J., & Prakken, H. (2011a). Revisiting preferences and argumentation, *Proceedings of the 22nd International Joint Conference on Artificial Intelligence*. AAAI Press,1021-1026.

Modgil, S. J., & Prakken, H. (2011b). A general account of argumentation with preferences. *Artificial Intelligence, 195*, 361-397.

Prakken, H. (2005). A study of accrual of arguments, with applications to evidential reasoning. *Proceedings of the Tenth In.ternational Conference on Artificial Intelligence and Law*, Bologna. New York:ACM Press, 85-94.

Prakken, H. (2010). An abstract framework for argumentation with structured arguments. *Argument and Computation, 1(2)*, 93-124.

Thomas, S. N. (1986). *Practical Reasoning in Natural Language, Third edition*. Prentice-hall in Englewood Cliffs..

Verheij, B. (1995). Accrual of arguments in defeasible argumentation. *Proceedings of the Second Dutch/German Workshop on Nonmonotonic Reasoning*, Utrecht, 217-224.

Vorobej, M. (1995). Hybrid Arguments. *Informal Logic, 17*, 289-296.

Walton, D.N. (1996). *Argument Structure: A pragmatic Theory*. Toronto, University of Toronto Press.

Walton, D.N. (2006). *Fundamentals of Critical Argumentation*, University of Cambridge Press, 139-149.

Chapter 2

A Theoretical Framework to Support Argumentation on Information Sources

Cristiano Castelfranchi, Rino Falcone, & Alessandro Sapienza

Institute for Cognitive Sciences and Technologies, ISTC-CNR, Rome, Italy
{cristiano.castelfranchi, rino.falcone, alessandro.sapienza}@istc.cnr.it

Abstract. Given many sources of information asserting different things, deciding to accept or not a belief can become a complicated task. Seeing the problem from the source's point of view, it would be interesting to understand what are the variables influencing the final outcome, in order to manipulate them in the most convenient way: it could be easiest to attack or support another source rather than just report an information. The basic idea in this work is to propose a novel systematic approach to beliefs' argumentation. Exploiting the concept of trust, we designed and implemented a theoretical and computational model that identifies all the variables playing a role when an agent has to evaluate whether to accept or not a given belief. According to us, once identified these factors, all the argumentations to support or to deny the belief have to deal with them. Then we present a list of some possible argumentative manoeuvres to show how the framework works.

1. Introduction

Trust is a complex and *recursive* beliefs structure [Castelfranchi and Falcone, 1998, Falcone and Castelfranchi, 2001]. I believe that P (a given fact, proposition), but if and how much I believe in it depends on the origin of this knowledge, on its 'sources'. How did I receive this information to be believed or rejected?
The basic principle of believing or not that P, are:
- the more credible/trustworthy the source F, the more sure I am about P (degree/strength of the belief);

- the many the converging sources the more sure I am;
- if the degree of credibility of the belief is higher than my subjective 'threshold' (which depends on personality, domain, risks, ..) than I will believe that P.

However, also the structure on the 'sources' and of their trustworthiness is just a belief structure. For example:

(a) I do believe that the source of P is F' (John, my eyes, ...) (and I am more or less sure of that);
(b) I do believe that I received and understood correctly that 'transmission' of information;
(c) I do believe that F' is a reliable source.

And again: on which basis do I entertain my beliefs (a) (b) and (c)? On the basis of a new edge of beliefs and sources.

For example, I believe in (a) because John (F") said me that was F' who said that P. Do I believe in F"? On the basis of which beliefs and sources?

I believe that (b) because I remember my perception process without disturbs and troubles.

I believe that (c) because I believe that F' is both competent, expert on P domain D' (and I believe that the domain of P is D'), and I believe that F' is sincere and honest (has no reasons for deceiving/harming me).

And so on: edge by edge; a recursive stratification of beliefs, their supports and sources, and the beliefs *about* these sources. Beliefs and meta-beliefs, layer by layer.

For example, on which basis do I believe that F' is competent? Because I had previous experience with F' in that D? Or because of her credits and titles? (and am I sure that they are true and good?) Or on the basis of her reputation? And is people opinion reliable? etc.

On which basis do I believe that F' is sincere and honest? Perhaps it is sincere but wrong; or perhaps it has some reason for deceiving me about that.

Considering that sometimes these beliefs are explicit, sometimes they are just potential, implicit, merely procedural: the way I use or rely on a given data. For example, the fact that John said to me that P can be just a memory image, a memory trace that I consider (use) as what happened (past) and what I have perceived. But it can become an explicit belief, especially if it is questioned, attacked: "On which base you believe so?".

Now like the beliefs supports other beliefs (making them credible), which support other beliefs, and so on; analogously the invalidation, revision, attack to a belief propagates layer by layer. If I induce you (by arguing) to abandon the belief B1 that was the main reason to believe B2, you have to revise, probably abandon, B2; and so on[1].

---

[1] Actually also attacking B1, which is the basis for believing that B2, is an attack to its 'source'. In fact the 'source' of a belief, its origin, cannot only be perception ("I saw it!"), or communication ("He said that..."), but also reasoning, inference: since B1 then B2.

One of the main manoeuvre for inducing you to abandon a given belief is attacking its sources or their credibility.

In this sense the argumentation approach represents a relevant way for evaluating the strength/weakness of the different information sources and of their trustworthiness.

2. Trust in Information Sources

For trusting information sources we have to trust the sources as competent and reliable in the domain of the specific information content.

Basically, agents have three main way to get evaluation on others' competence and reliability:
- (a) previous *Direct Experience* with F (how F performed in the past interactions) on that specific information content, or better "memory" about, and the adjustment that we have made of our evaluation of F in several interaction, and possible successes or failures relying on its information;
- (b) *Recommendations* (other individuals Z reporting their direct experience and evaluation about F) or Reputation (the shared general opinion of others about F) on that specific information content; (Yolum and Singh, 2003; Conte and Paolucci, 2002; Sabater-Mir, 2003; Sabater-Mir and Sierra, 2001; Jiang et al, 2013);
- (c) *Categorization* of F (it is assumed that a source can be categorized and that it is known this category), exploiting inference and reasoning:
    inheritance from classes or groups to which Z belongs (as a good "exemplar");
    analogy: Z is (as for that) like Y, Y is good for, then Z too is good for;
    analogy on the task: Z is good/reliable for P he should be good also for P', since P and P' are very similar. (In any case: how much do I trust my reasoning ability?).

On this basis it is possible to establish the competence/reliability of F on the specific information content [Falcone et al, 2013, Burnett et al, 2010].

The two faces of F's trustworthiness (competence and reliability) are relatively independent[2]; we will not treat them as such, by simplifying these complex components in just one quantitative parameter: F's estimated trustworthiness; by combining competence and reliability.

---

[2]Actually they are not fully independent. For example, F might be tempted to lie to me if/when is not so competent on providing good products: he has more motives for fudging me.

Moreover, information sources own and give us a specific information that they know/believe; but believing something is not a yes/no status. They can be more or less convinced (on the basis of their evidences, sources, reasoning). Thus a good source might inform us not only about P, but also about its *degree of certainty about P*, its trust in the truth of P. For example: "It is absolutely sure that P", "Probably P", "It is frequent that P", "It might be that P", and so on.

Of course there are more sophisticated meta-trust dimensions like: how much am I sure, confident, in F's evaluation of the probability of the event or in its subjective certainty?[3] Is F not sincere? Or not so self-confident and good evaluator?

For the moment, we put aside that dimension of how much meta-trust we have in the provided degree of credibility. We will just combine the provided certainty of P with the reliability of F as source. It in fact makes a difference if an excellent or a mediocre F says that the degree of certainty of P is 70%.

Another very relevant point, especially for information sources, is the following form of trust: the trust we have that the information under analysis derives from that specific source, how much we are sure about the "transmission", that the communication has been correct and working (and complete): there were interferences and alterations? Did I receive and understand it correctly? Is really F how is saying to be (Identity). Otherwise I cannot apply the first factor: F's credibility.

Let's simplify also these dimensions, and formalize just the *degree of trust that F is F*; that the F of that information (I have to decide whether believe or not) is actually F. In the WEB this is an imperative problem: the problem of the real *identity* of the F, and of the reliability of the signs of that identity, and of the communication.

These dimensions of trust on information sources (TIS) are quite independent on each other (and we will treat them as such); we have just to combine them and provide the appropriate dynamics. For example, what does it happen if a given very reliable source F' says that "it is sure that P", but I'm not sure at all that the information really comes from F' and I cannot ascertain that?

2.1 *Additional Problems and Dimensions*

We believe in a given datum on the basis of its origin, its source: perception? communication? inference? And so on.

(a) *The more reliable (trusted) the F the stronger the trust in P, the strength of the Belief that P.*

This is why it is very important to have a "memory" of the sources of our beliefs.

However, there is another fundamental principle of the degree of credibility of a given belief (its trustworthiness):

---

[3] In a sense it is a *transitivity principle* (Falcone and Castelfranchi, 2012): X trust Y, and Y trust Z; will X trust Z? Only if X trusts Y "as a good evaluator of Z and of that domain". Analogously here: will X trust Y because Y trusts Y? Only if X trust Y "as a good and reliable evaluator" of it-self.

(b) *The many the converging sources, the stronger our belief* (of course, if there are no correlations among the sources).

Thus we have the problem to combine different sources about P (that can be converging or diverging), taking into account their subjective degrees of certainty and their credibility in order to weigh the credibility of P, and have an incentive due to a large convergence of sources.

There might be different heuristics for dealing with contradictory information and sources. One (*prudent*) agent might adopt as assumption the worst hypothesis, the weaker degree of P; another (*optimistic*) agent, might choose the best, more favorable estimation; another agent might choose the most reliable source. We will formalize only one strategy: the weighing up and combination of the different strengths of the different sources.

2.2 Feedback on Source Credibility/TIS

We have to store the sources of our beliefs because, since we believe on the basis of source credibility, we have to be in condition to adjust such credibility, our TIS, on the basis of the result. If I believe that P on the basis of the source F1, and later I discover that P is false, that F1 was wrong or deceptive, I have to readjust my trust in F1, in order to be more prudent next time (with F1 or other similar sources). It is also the same also in case of positive confirmation[4].

However, remember that it is well known (Urbano et al, 2009) that the negative feedback (invalidation of TIS) is more effective and heavy than the positive one (confirmation). This asymmetry (the collapse of trust in case on negative experience versus the slow acquisition or increasing of trust) is not specific of trust and of TIS; it is -in our view- basically an effect of a general cognitive phenomenon. It is not an accident or weirdness if the disappointment of trust has much stronger (negative) impact than the (positive) impact of confirmation. It is just a sub-case of the general and fundamental asymmetry of negative vs. positive results, and more precisely of "losses" against "winnings": the well-known Prospect theory (Kahneman and Tversky, 1979). We do not evaluate in a symmetric way and on the basis of an "objective" value/quantity our progresses and acquisitions versus our failures and wastes, relatively to our "status quo". Losses (with the same "objective" value) are perceived and treated as much more severe: the curve of losses is convex and steep while that of winnings is concave. Analogously the urgency and pressure of the "avoidance" goals is greater than the impulse/strength of the achievement goals (Higgins, 1997). All this applies also to the slow increasing of trust and its fast decreasing; and to the subjective impact of trust disappointment (betrayal!) vs. trust confirmation. That's why usually we are prudent in deciding to trust somebody; in order do not expose us to disappointment and betrayals, and harms. However, also

---

[4] We can even memorize something that we reject. We do not believe to it but not necessarily we delete/forget it, and its source. This is for the same function: in case that information would result correct I have to revise my lack of trust in that source.

this is not always true; we have quite naive forms of trust just based on gregariousness and imitation, on sympathy and feelings, on the diffuse trust in that environment and group, etc. This also plays a crucial role in social networks on the web, in web recommendations, etc.

Moreover, according to our theory (Falcone and Castelfranchi, 2004) it is not always true that a bad/good result automatically entails the revision of TIS. It depends on the "causal attribution": has it been a fault/defect of F or an interference of the environment? The result might be bad although F's performance was his best. Let us put aside here the feedback effect and revision on TIS.

2.3 Plausibility: the Integration with Previous Knowledge

To believe something means not just to put it in a file in my mind; it means to *"integrate" it with my previous knowledge.* Knowledge must be at least non-contradictory, and possibly supported, justified: this explains that, and it is explained, supported, by these other facts/arguments. If there is *contradiction* I cannot believe P; either I have to reject P or I have to revise my previous beliefs in order to coherently introduce P. It depends on the strength of the new information (its credibility, due to its sources) and on the number and strength of the internal oppositions: the value of the contradictory previous beliefs, and the extension and cost of the required revision. That is: *it is not enough that the candidate belief that P be well supported and highly credible*; is there an epistemic conflict? Is it "implausible" to me? Are there antagonistic beliefs? And which is their strength? The winner of the conflict will be the stronger "group" of beliefs. Even the information of a very credible source (like our own eyes) can be rejected!

3. Formalizing and Computing the Degree of Certainty as Trust in the Belief

As we have said, there is a confidence, a trust in the beliefs we have and on which we rely.

Suppose X is a cognitive agent, an agent who has beliefs and goals.
Given Bel_X, the set of the X's beliefs, then P is a belief of X if:

$$P \in Bel_X \tag{1}$$

The degree of subjective certainty or strength of the X's belief P corresponds with the X's trust about P, and we call it:

$Trust_X(P)$ (2)

With $0 \leq Trust_X(P) \leq 1$

In the case in which $Trust_X(P) \geq th_X$ then $P \in Bel_X$. Here th_X stands for an X's internal minimal acceptance value (threshold) that determines if the degree of trust in a given statement is enough to accept it as a belief.

3.1 Its Origin/Ground

Concerning a single information P, we have to consider n different sources asserting or denying P. The final value of $Trust_X(P)$ depends on X's trust (degree of belief) on P according to the information coming from the different sources F (F_1, .., F_n):

$TrustSource_X(F, P)$ (3)

with $0 \leq TrustSource_X(F, P) \leq 1$

In other words, we state that:

$Trust_X(P) = f_1(TrustSource_X(F_1, P), ..., TrustSource_X(F_n, P))$ (4)

Where n is the total number of sources relatively to P and f_1 is in general a function that preserves monotonicity and ranges in [0,1].

Then to compute X's trust value in P, X has to compose the different information coming from n sources in just one resulting factor. This composition follows the conceptual modeling described in the §2 where we have that $TrustSource_X(F, P)$ can be articulated in:

> X's degree of trust on F regarding the judgment of P. This value is built on the X's judgment about the different attitudes (both competence and reliability) of F as derived by the composition of three factors of different nature (direct experience, recommendation/reputation, and categorization). In practice the F's credibility about P according to X. We call this first factor as *MainTrust*:

$MainTrust_X(F, P)$ (5)

with $0 \leq MainTrust_X(F, P) \leq 1$

The F's degree of certainty about P: information sources may not only give the information but they also have (and may communicate) their certainty about that information; we are interested in this certainty; in particular we assume that X receives/perceives exactly the same value considered by F:

$Trust_F(P)$

    The X's degree of trust that P originates from F; X believes that P is from source F to the degree:

$Trust_X(Source(P) = F)$

    where the operator $Source(P)$ can return different sources (F_1, ... F_n). We are interested to test F as source of P.

    The fact that F is supporting P or is opposing to it (not P):

$Trust_F(P) \geq th_F$ or $Trust_F(P) \leq th_F$

$TrustSource_X(F, P) = f_2(MainTrust_X(F, P), Trust_F(P), Trust_X(Source(P) = F), th_F)$

    where f_2 ranges in [0,1].

3.2 A Modality of Computation

As specified in §2 the value of $MainTrust_X(F, P)$ is a function of:
Past interactions;
The category of membership;
Reputation.

    These dimensions are represented by fuzzy sets (see Figure 1 below): terrible, poor, mediocre, good, excellent. We then compose them into a single fuzzy set, considering a weight for each of these three parameters. Those weights are defined in range [0;10], with 0 meaning that the element has no importance in the evaluation and 10 meaning that it has the maximal importance.

    It is worth noting that the weight of experience has to be referred to a twofold meaning: it must take into account the numerosity of experiences (with their positive and negative values), but also the intrinsic value of experience for that subject.

    However, the fuzzy set in and by itself is not very useful: what interests us in the end is to have a plausibility range, which is representative of the expected value of $MainTrust_X(F, P)$.

    To get that, it is therefore necessary to apply a defuzzyfication method. Among the various possibilities (mean of maxima, mean of centers ...) we have chosen to use the *centroid method*, as we believed it can provide a good representation of the fuzzy set. The centroid method exploits the following formula:

$$k = \frac{\int_0^1 x f(x)\,dx}{\int_0^1 f(x)\,dx}$$

were f(x) is the fuzzy set function.

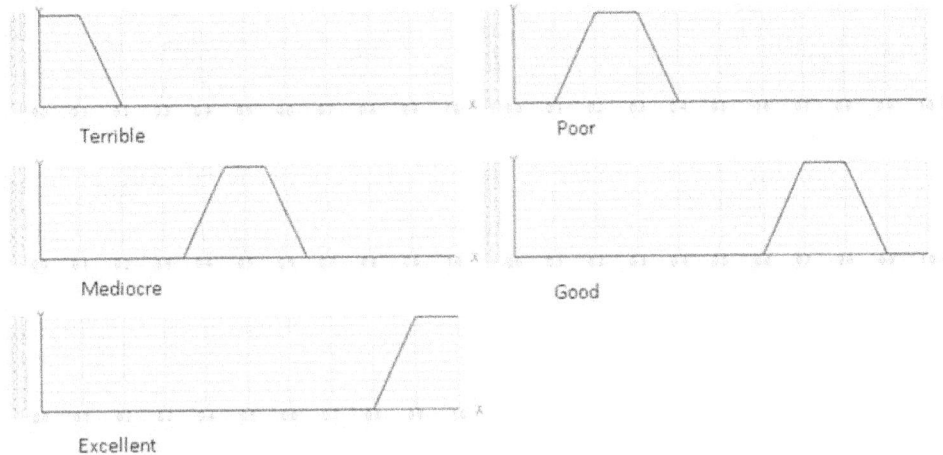

Figure 1. Representation of the five fuzzy sets

The value k, obtained in output, is equal to the abscissa of the gravity center of the fuzzy set.

This value is also associated with the variance, obtained by the formula:

$$\sigma^2 = \frac{\int_0^1 (x-k) f(x)\, dx}{\int_0^1 f(x)\, dx}$$

With these two values, we determine $MainTrust_x(F,P)$ as the interval [k-σ;k+ σ].

3.3 Trust Value and Sources' Aggregation

Putting aside mathematical formulations, in this part we intend to provide the workflow of the model, describing how input parameters influence and are linked to the output.

Once we get $MainTrust_x(F,P)$, we are able to determine the value of $TrustSource_x(F,P)$, that is the contribute that the source F gives to X. It will depends on:

1. $MainTrust_x(F,P)$: the trust evaluation of F according to X;
2. $Trust_F(P)$: how much P is supporting P;

3. $Trust_x(Source(P) = F)$: how much strong is the link between F and P;
4. th_F: the fact that F is supporting or not P.

Initially, the value of $TrustSource_x(F, P)$ is set to 0.5. This value has been chosen because of its particular meaning: it is neutral; it does not give any information about trusting or not F.

Then the other parameters will modulate it in a multiplicative manner, letting trust either increase or decrease.

Finally the trust evaluations, representing the informative contribute of each source, has to be aggregated, in order to provide the trustor with a single and expressive final trust evaluation.

Given that each trust evaluation has been properly smoothed and weighted according to the confidence that the trustor has on F and that F express on its own belief, we decided to express this summarizing value through the average of individual sources' trust evaluations.

The whole model can be summarized as in the following picture, that clearly shows which are inputs and outputs in each phase, starting from the source evaluation till the production of the final trust value.

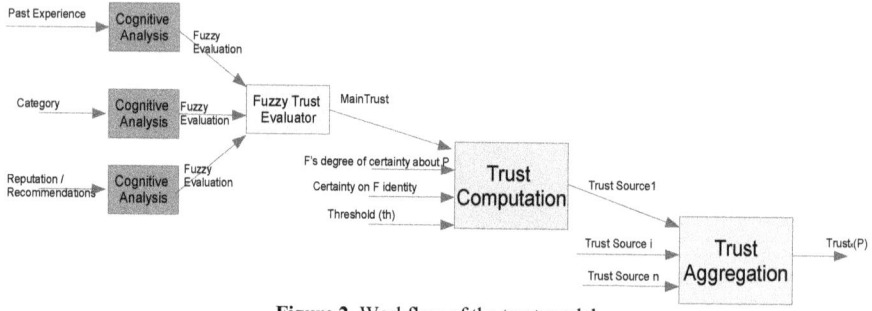

Figure 2. Workflow of the trust model

4. The Argumentation Framework

Other works deal with similar topics (Hughes et al, 2014) investigate how attacking the expertise and trustworthiness and supporting the one making the attacks influences source credibility.

(Dunbar *et al*, 2014) study credibility, persuasion processing cues and interactivity factors in websites of non-ideological groups, non-violent ideological groups, and violent ideological groups, using the elaboration likelihood model (ELM) as a theoretical framework.

What we want to do here is to provide a systematic method to approach information's source argumentation.

Starting from our previous analysis about trust in information sources, we would like to extend this approach to the argumentation about that information. In particular, we should consider the different sources for trusting information and analyse them in the perspective of finding arguments supporting or contradicting it.

Notice that these argumentation "moves" are not "collected" on the field, like in botanic, but derived in a principled way from the dimensions of the model. In fact we state that, once identified all the components that influence trust in a given information source, any kind of support/attack has to deal with these components.

Let's present these maneuvers in form of attack and discredit moves against F, based on the various origins on trust-beliefs about F: personal experience, reputation, reasoning, and so on.

Of course, each move has its correspondent positive version in terms of accredit and reinforcement of F's credibility. Our analysis does not claim to be complete; it just to provide hints and examples for a method.

4.1 Attacks on X's Previous Experience and Interactions with F

X can trust/distrust an information P ($Trust_X(P)$) because its direct experience with F (source of P) was positive/negative. We can find arguments on this dimension in the following ways:

- it could be argued in favour (or against) the number of interactions X had with F about P (or with the information class to which P belongs): Are they enough or not enough? For that kind of information (P is in a specific class of information) it is necessary less (or more) interactions than the ones X experienced with F (*argumentation about the amount of interactions*).
- *argumentation about the typology of interactions*:
  1a) "OK, you had good experiences with F, but in those circumstances he had no reasons for deceiving you, but now he is involved, has personal interests in your believing ("pro domo mea")". (*Attack on F's willingness, disposition, and honesty*).
  1b) "OK, you had good experiences with F, but they were in a completely different domain! Are you sure that he is equally expert on P?" (*Attack to F's competence*).
  1c) "OK, you had good experiences with F, but they were in a completely different context! Are you sure that in a different environmental context he is equally able and willing on P?" (*Attack to the role of context on F's performance*).
  1d) "Are you sure that you have a correct memory of that? Or have you adjusted your memories by forgetting or putting aside something that would be relevant?" (*Attack to the X's memory about the F's performance*).

1e) "That's strange; your personal opinion about F is very different from what all the others think" (*Attack to the X's evaluation on F with respect to the F's reputation*).
1f) "As usual: you are too naive and generous(/ungenerous) with the others; you believes that everybody is honest(/dishonest) and serious (not serious)" (*Attack to the general ability to evaluate by X*).

Several of these maneuvers actually are meta-attacks: I attack *you* as a good source of your own beliefs/evaluations about F (based on personal experience).

Of course, some of these maneuvers are quite risky for the arguer; in fact they try to destroy several supporting beliefs of X, even on himself. The many the integrated and supporting beliefs of X that should be revised or abandoned, the more costly this readjustment will be, the more demanding for X to believe to the argument of the arguer, the more "resistant" he will be, with a higher probability to reject not only of the argument but also the arguer. The arguer's trustworthiness should be very high to force X to revise so many of his beliefs. Thus, the arguer has to calculate this balance and risk.

4.2 Attacks on F's Reputation and Recommendations From Others

Clearly, all the maneuvers that we are illustrating about F can be recursively applied to the sources of information about F (recommendations, reputation, ..). One can destroy F's reputation just by destroying its favorable sources. And so on.

Let's see some specific moves:

2a) "Are you sure that they are good evaluators in this case/domain/context?" For example: "It is true, his reputation is excellent, but on other domains; he is an excellent teacher but research is not the same thing"
2b) "Are you sure that they are disinterested, not involved or corrupted?"
2d) "Do you remember well what they told you? Or do you adjust your memory as convenient?" (like 1d).

4.3 Attacks on F's Inherited Trustworthiness

This kind of arguments should consider the domain of trustworthiness gained form the source on different potential dimensions.

3a) "Are you sure that F is an medical doctor?"
3b) "Are you sure that F as doctor has the right competence for that advice?"

And so on.

4.4 F's Certainty (Perceived by X)

"Did X well understand what F said?" Of course, what it is possible to question here is not the objective certainty of F regarding P, but the perceived value of certainty from X's point of view: this certainty is questionable, since it depends on X's personal interpretation of F's words.

4.5 Source's Identity

This element has a twofold meaning:

1. Am I sure that F is the source of P?
2. Am I sure that F is what it is saying to be?

In the first case the problem is inherent to X's memory: is it remembering right? This sight is really important in the verbal interaction, but it looses its relevance in the web environment: if I'm not sure, I can always check (with a minimum cost to retrieve this information).

Actually in the web context the second case has a higher weight. In fact it is really easy, hiding behind a screen, to impersonate another person, creating fictitious characters or stealing someone else's identity.

To attack X's certainty about F's identity has a very strong effect: what is the point of evaluation what F is saying if I have no idea about who is F?

5. Conclusion

The two main achievement of this work are:
1. given the complex nature of trust on information sources, we wanted to provide a detailed theoretical and computational model to deal with it, also on the basis of our previous works [Castelfranchi et al, 2003; Castelfranchi et al, 2014];

on the base of this analysis, we wanted to exploit that model to create an argumentation framework.

This second point results particularly useful from the source's point of view. In fact we state that any possible argumentation attack or support to a belief has to act on one or more of the presented features. Knowing that represents a significant plus for an arguer.

Finally, we presented a series of possible form of attack against a source F, first basing on the various origins of trust-beliefs about F (personal experience, reputation, reasoning, and so on), then about what it is asserting, its self-confidence about it and its own identity.

It is worth noting that each move has its correspondent positive version in terms of accredit and reinforcement of F's credibility. Our analysis does not claim to be complete; we just intend to provide hints and examples to show the effectiveness of the proposed model.

Acknowledgments

This work is partially supported both by the Project PRISMA (PiattafoRme cloud Interoperabili per SMArt-government; Cod. PON04a2 A) funded by the Italian Program for Research and Innovation (Programma Operativo Nazionale Ricerca e Competitività 2007-2013) and by the project CLARA—CLoud plAtform and smart underground imaging for natural Risk Assessment, funded by the Italian Ministry of Education, University and Research (MIUR-PON).

References

Burnett, C., Norman, T., and Sycara, K. 2010. Bootstrapping trust evaluations through stereotypes. In Proceedings of the 9th International Conference on Autonomous Agents and Multiagent Systems (AAMAS'10). 241248.

Castelfranchi C. and Falcone R.. "Principles of trust for MAS: Cognitive anatomy, social importance, and quantification." *Multi Agent Systems, 1998. Proceedings. International Conference on.* IEEE, 1998.

Castelfranchi, C., Falcone R., Pezzulo, (2003) Trust in Information Sources as a Source for Trust: A Fuzzy Approach, Proceedings of the Second International Joint Conference on Autonomous Agents and Multiagent Systems (AAMAS-03) Melburne (Australia), 14-18 July, ACM Press, pp.89-96.

Castelfranchi C., Falcone R., Sapienza A., Information sources: Trust and meta-trust dimensions . CEUR Workshop Proceedings 2014 (In press)

Conte R., and Paolucci M., 2002, Reputation in artificial societies. Social beliefs for social order. Boston: Kluwer Academic Publishers.

Dunbar, N. E., Connelly, S., Jensen, M. L., Adame, B. J., Rozzell, B., Griffith, J. A. and Dan O'Hair, H. (2014), Fear Appeals, Message Processing Cues, and Credibility in the Websites of Violent, Ideological, and Nonideological Groups. Journal of Computer-Mediated Communication, 19: 871–889. doi: 10.1111/jcc4.12083

Falcone, R., & Castelfranchi, C. (2001). Social trust: A cognitive approach. In Trust and deception in virtual societies (pp. 55-90). Springer Netherlands.

Falcone R., Castelfranchi, C. (2004), Trust Dynamics: How Trust is influenced by direct experiences and by Trust itself; Proceedings of the 3rd International Conference on Autonomous Agents and Multi-Agent Systems (AAMAS-04), New York, 19-23 July 2004, ACM-ISBN 1-58113-864-4, pages 740-747.

Falcone R., Castelfranchi C., Trust and Transitivity: How trust-transfer works, 10th International Conference on Practical Applications of Agents and Multi-Agent Systems, University of Salamanca (Spain)28-30th March, 2012.

Falcone R., Piunti, M., Venanzi, M., Castelfranchi C., (2013), From Manifesta to Krypta: The Relevance of Categories for Trusting Others, in R. Falcone and M. Singh (Eds.) Trust in Multiagent Systems, ACM Transaction on Intelligent Systems and Technology, Volume 4 Issue 2, March 2013

Higgins, E. T. (1997). Beyond pleasure and pain. American Psychologist, 52, 1280-1300.

Hughes, M. G., Griffith, J. A., Zeni, T. A., Arsenault, M. L., Cooper, O. D., Johnson, G., Hardy, J. H., Connelly, S. and Mumford, M. D. (2014), Discrediting in a Message Board Forum: The Effects of Social Support and Attacks on Expertise and Trustworthiness. Journal of Computer-Mediated Communication, 19: 325–341. doi: 10.1111/jcc4.12077

S. Jiang, J. Zhang, and Y.S. Ong. An evolutionary model for constructing robust trust networks. In Proceedings of the 12th International Conference on Autonomous Agents and Multiagent Systems (AAMAS), 2013.

Kahneman, D., & Tversky, A. (1979). Prospect Theory: An Analysis of Decision Under Risk. Econometrica, XLVII, 263-291.

Sabater-Mir J., Sierra C., (2001), Regret: a reputation model for gregarious societies. In 4th Workshop on Deception and Fraud in Agent Societies (pp. 61-70). Montreal, Canada.

Sabater-Mir, J. 2003. Trust and reputation for agent societies. Ph.D. thesis, Universitat Autonoma de Barcelona.

Joana Urbano, Ana Paula Rocha, and Eugnio Oliveira, Computing Con_dence Values: Does Trust Dynamics Matter? In L. Sabra Lopes et al. (Eds.): EPIA 2009, LNAI 5816, pp. 520-531, 2009, Springer.

Yolum, P. and Singh, M. P. 2003. Emergent properties of referral systems. In Proceedings of the 2nd International Joint Conference on Autonomous Agents and MultiAgent Systems (AAMAS'03).

Chapter 3

On the Rationality of Argumentative Decisions

Fabio Paglieri

Goal-Oriented Agents Lab (GOAL), Istituto di Scienze e Tecnologie della Cognizione, Consiglio Nazionale delle Ricerche (ISTC-CNR), Via San Martino della Battaglia 44, 00185 Roma, Italy.
fabio.paglieri@istc.cnr.it

Abstract. This paper summarizes the basic assumptions of a decision theoretic approach to argumentation, as well as some preliminary empirical findings based on that view. The relative neglect for decision making in argumentation theory is discussed, and the approach is defended against the charge of being merely descriptive. In contrast, it is shown that considering arguments as the product of decisions brings into play various normative models of rational choice. This presents argumentation theory with a novel challenge: how to reconcile strategic rationality with other normative constraints, such as inferential validity and dialectical appropriateness? It is suggested that strategic considerations should be included, rather than excluded, from the evaluation of argument quality, and this position is put in contact with the growing interest for virtue theory in argumentation studies.

1. Introduction

When we argue, we make several decisions, sometimes without even realizing it: we choose whether to enter the argument or not, what arguments to use and how to present them, how to react to the arguments of the counterpart, how to respond to challenges and objections, how to solve potential ambiguities, when and how to end the argument, and so forth. In fact, argumentation can be seen as the result of a complicated decision-making process, or, more exactly, as the interaction of multiple decision-making processes performed by autonomous agents.

In spite of its obvious relevance in everyday argumentation, decision-making has been taken for granted rather than explored in argumentation theories, with few exceptions (the work of Dale Hample and colleagues being one of the most notable cases; see Hample, 2005 for a review). This neglect originates from an insistence on what is the right move in a given argumentative situation, and not on how the subject may decide to opt (or not) for that move. It is not that argumentation theorists are unaware of argumentative decisions, of course – as arguers, if nothing else, they are bound to be familiar with those. They just do not see it as their business to produce a theory of such decisions.

This is certainly true for those approaches that concentrate mostly on argumentation as reasoning and arguments as valid inference schemes (these include, among others, Toulmin, 1958; Walton, 1996; Johnson, 2000, and most of the work done in informal logic).[1] Whatever decisions prompted the subject to apply a certain reasoning pattern or inference scheme are typically beside the point in this line of research, since what ultimately matters is the validity (or lack thereof) of the resulting argument. The fact that the concept of "validity" comes in many guises in argumentation theories, ranging from deductive validity to presumptive strength, does not change the perceived irrelevance of the decision-making process that produced a given argument, be it a silent piece of reasoning or an uttered statement.

The same is true for the degree of "dialecticity" (or "dialogicity", for those who consider this distinction relevant; see Walton & Godden, 2007) of a given approach: even if most theories make specific provisions to account for several dialectic features of argumentation, these are analyzed only insofar as they bear on the acceptability of the resulting argument. For instance, in commitment-based analyses of argumentation (Walton & Krabbe, 1995, being a case in point), what matters is what arguers can be reasonably held accountable for, regardless of what made them decide to incur such dialogical obligations (for further critical discussion on that project, see Paglieri, 2010). The virtues or vices of the end product do not depend on the underlying decision-making process, so theories focused on argument-as-product can safely ignore argumentative decisions. Even more crucially, insofar as argumentation is considered a form of reasoning, it is likely that whatever process is responsible for it, it is not a type of decision-making: after all, reasoning results in beliefs (possibly beliefs about what to do), and beliefs are often thought as being formed outside of the volitional control of the believer – what Woods aptly named "doxastic irresistibility" (2005, p. 755).[2]

According to a popular way of carving up the rich field of argumentation studies, argument-as-product approaches are complemented by theories of argumentation-as-process (Reed & Walton, 2003; a similar distinction with different labels is found in O'Keefe, 1977, and Habermas, 1984, among others). It would seem natural for the latter to provide an in-depth analysis of argumentative decisions: seen as an activity, argumentation clearly includes and even requires a series of decisions on what

---

[1] Incidentally, in their formative years informal logic and argumentation theory had historical reasons for moving away from formal theories in general, including rational choice theory, and these reasons were valid at the time (Toulimin, 1958 still provides a powerful illustration). However, formal logic has undergone a profound transformation in recent decades, so nowadays I concur with Johan van Benthem, when he notes that «the role of richer argument schemata and of procedural aspects of reasoning (…) are in fact shared interests, making logic and argumentation theory allies rather than rivals» (2009, p. 13).

[2] The opposite view, doxastic voluntarism, has its fair share of defenders (e.g. Steup, 2000; Wansing, 2000, 2002, 2006; Ginet, 2001; Ryan, 2003), but is by far a minority position in the epistemology of belief – and rightly so, in my view.

argumentative moves to make in the course of dialogical interchange. However, theories of argumentation-as-process merely countenance but do not explain argumentative decisions, since their main focus is typically normative rather than descriptive. An obvious example is pragma-dialectics (for the most recent version, see van Eemeren & Grootendorst, 2004): a critical discussion is defined as a purposive activity aimed at solving a difference of opinion, where arguers are free to decide on their moves within the boundaries of rational rules and shared standpoints, each of them striving to strike a balance between effectiveness and reasonableness (strategic maneuvering; see van Eemeren & Houtlosser, 2002, 2006). However, pragma-dialecticians are interested to identify the structure and rules that arguers ought to follow to be rational, not to discuss what motivates their decisions between rationally acceptable options – aside from a generic commitment to effectiveness in strategic maneuvering. As a result, the factors affecting the arguers' decisions in a critical discussion are left mostly in the background, whereas full prominence is given to the dialectical obligations they incur by making such decisions.

In contrast with this traditional lack of scholarly interest for argumentative decisions,[3] recent studies have proposed to analyze argumentation as a decision-making process, both theoretically (Paglieri, 2009, 2013; Paglieri & Castelfranchi, 2010a) and experimentally (Cionea et al., 2011; Hample et al., 2011). In what follows I will first recap the main results of this line of research, and then address a criticism that is likely to be raised against it: namely, the alleged *lack of normative concerns* expressed by a decision-theoretic approach to argumentation. I will claim that this criticism is ultimately misguided, since looking at arguments as decisional processes does not entail abandoning a normative perspective, in favour of a merely descriptive approach. On the contrary, it raises a host of new normative questions, regarding the interplay between inferential validity, dialectical appropriateness, and strategic rationality.

2. Some Results on Argumentative Decisions

In presenting a typology of argumentative decisions, it is useful to follow the typical chronological order in which the arguer has to face these decisions. This criterion is by no means the only possible one, and the resulting taxonomy is not necessarily intended to be exhaustive. But it does provide a convenient starting point, both by giving some order to the discussion, and by helping to better illustrate what an argumentative decision looks like "in real life". In light of these considerations, what

---

[3] It is worth noting that such neglect is most apparent in philosophical studies of argumentation, whereas in computational models of argument it is frequent to encounter some attention for how argumentative moves are selected (see for instance Amgoud & Maudet, 2002; Karunatillake & Jennings, 2005; Riveret et al., 2008; Rahwan & Larson, 2009). However, none of these approaches offers a comprehensive decision-theoretic perspective on argumentation, as we discussed elsewhere (Paglieri & Castelfranchi, 2010a).

follows is meant as a first, tentative process-based taxonomy of argumentative decisions (discussed in greater detail elsewhere; see Paglieri, 2009, 2013):[4]

(I) *Argument engagement*: the decision to enter an argument or not, either by proposing one or by accepting to be drawn into one by the counterpart. Considering engagement as a decision implies acknowledging that arguing is not always the best option, and sometimes it is actually the worst (Martin & Scheerhorn, 1985; Hample & Benoit, 1999; Cohen, 2005; Goodwin, 2005, 2007; Paglieri, 2009). More generally, the strategic considerations that are relevant in choosing whether to argue or not are best understood in terms of costs and benefits, as exemplified by various contributions in Artificial Intelligence (Amgoud & Maudet, 2002; Karunatillake & Jennings, 2005; Riveret et al., 2008; Rahwan & Larson, 2009; Paglieri & Castelfranchi, 2010a).[5]

(II) *Argument editing*: all decisions concerning what arguments to use (selection) and how to present them to the audience (presentation), in order to maximize their intended effects (Hample & Dallinger, 1990, 1992; Hample, 2005; Hample et al., 2009). These argumentative decisions also have obvious relevance in rhetoric, and in fact the five canons of Western classical rhetoric (*inventio, dispositio, elocutio, memoria, pronuntiatio*) can be seen as a practical way of carving up different sub-decisions concerning effective argument presentation.

(III) *Argument timing*: the decision on when it is time to speak and when it is time to listen. An appropriate timing of one's argumentative contribution and an awareness about the optimal length of one's speech are essential elements for making an effective argument in almost every dialogical context, including one-way presentations in front of an audience.

(IV) *Argument interpretation*: if there are ambiguities in what the counterpart is saying, the decision on whether to criticize them, ask for more clarity or additional information, or solve them autonomously – and if so, favoring what interpretation, on what grounds, and to what ends? A well-studied case of argument interpretation concerns enthymemes, but most theories of enthymemes focus on what is the normatively correct/legitimate reconstruction of the argument. To highlight the

---

[4] In this taxonomy, and more generally throughout the paper, the word "argument" is used with two different meanings: "argument" as a prolonged dialogical exchange between two or more parties, possibly (but not necessarily) of an adversarial and controversial nature, and "argument" as a series of premises in support of a certain conclusion, typically advanced by a single speaker. Both meanings are standard and non-technical, so I see no reason to introduce any specific notation to distinguish when the word "argument" is used to refer to one or another: the context will suffice to allow proper disambiguation. Quite clearly, for instance, the decision to enter or terminate an argument refers to the first meaning (roughly, it means initiating or concluding an argumentative discussion with another party), whereas deciding what arguments to use in a discussion manifestly refers to the second meaning (to wit, choosing what premises to offer in support of one's position).

[5] Costs and benefits are crucial also in relevance theory (Sperber & Wilson, 1986; Wilson & Sperber, 2002), but with several important differences with respect to the decision-theoretic approach outlined in this paper. The point is discussed at some length elsewhere (Paglieri, 2013), hence I will not dwell further on it here.

arguer's underlying decisions, pragmatic approaches such as relevance theory (Sperber & Wilson, 1986; Wilson & Sperber, 2002) are more useful (for additional details on this point, see Paglieri, 2007; Paglieri & Woods, 2011a, b).

(V) *Argument reaction*: decisions concerning whether to accept or challenge an argument, an objection, or a counter-argument raised by the counterpart. This is another area that has been so far dominated by normative concerns: the widely received wisdom in argumentation theories is that arguments should be challenged and critical questions should be asked, whenever appropriate, but there has been little consideration on what reasons (other than being right) might guide this choice (see Gilbert, 1997, for some in-depth discussion of these issues, as well as a critique of the enduring lack of attention they suffered in argumentation theories).

(VI) *Argument termination*: the decision on when and how to end an ongoing argument. Clearly the arguer cannot unilaterally "decide" to win the argument (or to reach whatever goal s/he pursued by arguing), since, for this to happen, the agreement of the counterpart and/or the satisfaction of some objective criteria are required. But each arguer can and do decide whether to let the other win by conceding the point, or shelve/postpone the argument, or move on to other matters, or some other way of terminating the argumentative exchange for the time being (for some preliminary findings on this point, see Benoit & Benoit, 1987, 1990; Vuchinich, 1990; Hicks, 1991; Hample et al., 1999).

In spite of the preliminary nature of this taxonomy, there are already some empirical evidence of its usefulness, in particular concerning argument engagement and argument editing. On argument engagement, it has been empirically verified that the intention to participate in an argumentative episode is predicted more effectively by strategic considerations, in interaction with social context (Cionea et al., 2011; Hample et al., 2011), than by personality traits, such as argumentativeness and verbal aggressiveness (Infante & Rancer, 1982; Infante & Wigley, 1986). On argument editing, several studies have shown that the choice of what arguments to use and how to present them is finely tuned to the specific goals of the arguers in a given context of interaction, rather than descending from their overall character dispositions (Hample & Dallinger, 1990, 1992; Hample, 2005; Hample et al., 2009).

These results encourage further efforts at empirical verification of predictions based on a decision-theoretic approach to argumentation. For instance, it has recently been proposed (Paglieri, 2013) that argument termination is modulated by length of interaction, and in particular that arguers increase their propensity to stop arguing as a function of how much time they already spent doing so. Appropriate experimental manipulations would allow not only to verify or falsify this general prediction, but also to establish what factors determine such propensity to "keep it brief": an increase in the costs of the argument ("Enough, I am tired of wasting time with you!"), a higher likelihood of undesirable side-effects ("Let's argue no more, lest we end up fighting!"), a stronger skepticism on the possibility of reaching a satisfactory conclusion ("If we haven't agreed so far, why keep arguing any longer?"), or some combinations of the above. It is not hard to imagine suitable methods to test such hypotheses, as well as others of the same nature – provided we accept to look at arguments (also) as decision-making processes.

3. Strategic Rationality, Inferential Validity, and Dialectical Appropriateness

Regardless of the promising empirical results summarized in the previous section, argumentative decisions may still be regarded as alien to the objectives and methods of argumentation theories. Even granting that such decisions are a worthy object of study, it does not follow that the task of studying them should be appointed to argumentation scholars. Indeed, precisely because argumentation theories are eminently normative (they deal with how a rational agent *ought* to argue) and not descriptive (they do not care much for how real agents in fact argue, except to look for deviations from normative standards), it is debatable that they should pay special attention to purely empirical facts, such as the choices that lead people to argue in a certain manner.

In answering this objection, one can either negate that argumentation theories should confine themselves to normative concerns, or prove that the study of argumentative decisions is not confined to a merely descriptive level. Whereas I confess deep misgivings on a purely normative view of argumentation theories,[6] this is not a point that I want to press here, because I see the issue as being mostly "political": that is, it is a matter of establishing the proper object of analysis of argumentation theories, and this is bound to require a wide consensus and thus engender vibrant debate. For the purpose of this paper, it is both easier and more interesting to demonstrate the flaw in the second premise of this criticism, to wit, the idea that normative concerns are excluded from the study of argumentative decisions. Nothing could be further from the truth.

In fact, as soon as we look at argumentative process as decision-making processes, we are able to make use of several well developed theories of rational choice, such as expected utility maximization, satisficing procedures *a la* Simon, prospect theory, ecological heuristics, and many more. Each of these normative theories can be applied to argumentative decisions without much tinkering, thus producing clear intuitions on what should count as a *rational* argumentative choice, and from what general normative principles this judgement is derived. Moreover, if

---

[6] My misgivings depend on what I see as a confusion between two different pairs of opposites: "normative vs descriptive", and "general vs particular". While I take it as a given that argumentation theories, *qua* theories, should aim to extrapolate general models of argumentative processes from a myriad of particular concrete instances, instead of just collecting facts, I do not see why such theories should necessarily be only, or even mainly, of normative nature. Clearly, defining standards of correctness for argumentation is a noble endeavor. But it is no less noble to account, with appropriate generality, for what mechanisms and processes determine the shape of real-life arguments, as they are observed empirically. Thus I find the identity "argumentation theories = normative approaches" to be misleading and limiting: at best, it registers an objective predominance of normative concerns in the extant literature, but this is the fruit of historical accident, not of theoretical necessity.

emphasis is to be given to strategic interaction between arguers, it is straightforward to adopt several well-tested tools for the analysis of decision-making in social contexts, such as game theory (for instances of this approach, see Paglieri & Castelfranchi, 2005; Riveret et al., 2008; Rahwan e Larson, 2009). Nor it is problematic to adapt the analysis to more realistic scenarios, by using models of bounded rationality to study argumentative decisions: a possibility already advocated by scholars who insist on the importance of studying argumentation with an eye to the cognitive boundaries of arguers (e.g., Wilson & Sperber, 2002; Gabbay & Woods, 2003; Paglieri, 2009; Paglieri & Castelfranchi, 2010a, b; Paglieri & Woods, 2011a). In sum, far from relinquishing normative concerns in the study of argumentation, a decision-theoretic perspective on argumentative processes offers for free many powerful tools for defining and assessing the rationality (or lack thereof) of arguers.

Clearly, here we are referring to *strategic rationality*, that is, what is rational to do, given a certain set of preferences, beliefs, and current external conditions. Admittedly, this is very different from the kind of rationality usually discussed in argumentation theories – but now we can start appreciating that this is due to a staunch refusal of looking at argumentative decisions, rather than to a legitimate predilection for a normative stance. Indeed, argumentation theories typically focus on the *inferential validity* of arguments (whether the conclusion follows from its premises, and with which strength) and the *dialectical appropriateness* of argumentative moves (under what conditions an arguer is permitted and/or required to undertake certain dialogical acts, given a shared set of norms on how to conduct a rational discussion).

These normative concerns are absolutely relevant, but not necessarily antagonistic with respect to the strategic rationality of argumentative decisions made by speakers. On the contrary, it is quite clear that the strategic value of an argumentative move is at least partially *independent* from both the inferential validity of the argument on which it is based, and the dialectical legitimacy that authorizes its use in that context. For instance, as far as it is reasonable to expect the counterpart not to notice a logical or dialectical infraction, and such infraction allows the arguer to pursue his/her agenda optimally, then it is strategically rational to commit that infraction, even if the resulting argument is inferentially invalid and dialectically incorrect. Conversely, a logically excellent argument might turn out to be strategically useless (e.g., if the counterpart is unable to follow its inferential steps) or even damaging (e.g., if it is bound to lead to violent escalation of the social conflict – being right did not do much good to the Talking Cricket in *Pinocchio*, as readers may recall); the same applies to an argumentative move that is dialectically appropriate but practically ineffective, or worse (e.g., legitimate criticism may be construed as a personal aggression, which is something to be avoided in various contexts).

4. Does Strategic Rationality Matter for Argument Evaluation?

Even if it makes sense to speak of strategic rationality with respect to argumentative decisions, and even if this dimension cannot be reduced to other normative concerns, one might still object that strategic rationality should not be a dominant factor in establishing what is the right way of arguing. On the contrary, a staunch opponent to the study of argumentative decisions might insist that strategic rationality should remain *subordinated* to other, more relevant normative standards – again, inferential validity and dialectical appropriateness. In this view, an argumentative decision, in order to be rational, *has* to produce a valid argument and/or an appropriate move, regardless of what other practical goals it might satisfy. Validity and appropriateness are necessary (although not sufficient) conditions for the overall rationality of argumentative decisions. In contrast, strategic rationality really matters, normatively speaking, only to choose between equally valid and equally appropriate argumentative options. Such role is, at best, ancillary; at worst, it is irrelevant, when there is no more than one maximally valid and optimally appropriate option to select from.

This restrictive view of the role of strategic rationality in argumentation may seem to have a certain intuitive appeal. After all, the fact that there are practical reasons (e.g., swaying the audience, or avoiding harsh confrontation) to prefer a sub-optimal argumentative option (e.g., an invalid argument) does not make that option any better. Put it more simply, an invalid or inappropriate argument does not become valid or appropriate only because you use it smartly to achieve your practical goals, and get away with it. This is of course correct. However, it misses the point, which is *not* to reduce validity or appropriateness to strategic rationality, but rather to claim that the latter can have equal or even higher normative status than the former, when it comes to assess the rationality of a certain argumentative behavior. The point is not to conflate these different dimensions of rationality, but rather to put all of them on equal footing. Just as inferential validity can overrule strategic rationality in certain contexts (e.g., in scientific debate), the opposite can also happen.

Cases where strategic considerations do (and ought to) outweigh other normative concerns abound in political discourse. Imagine for instance that there are valid economic reasons to cut down on salaries of public employees, because this will ultimately result in a benefit for the economy as a whole and even public employees will be better off in the end, in spite of their reduced wages. However, the reasoning required to grasp this chain of economical causes and effects happens to be well beyond the understanding of laymen, and getting involved in technicalities would only muddle the issue and engender suspicion in the public eye. Hence, the political party promoting this reform opts for packaging the proposal as a fight against the rampant corruption, widespread inefficiency, and undeserved privileges of public servants. This succeeds in creating strong support for the proposal in the general public, although it makes public employees dead against these cuts (but this would have happened anyway, most likely). As a result of this argumentative strategy, the reform is voted and comes into effect: luckily, the expected benefits ensue, and in the long run everybody is happy.

Now, assuming side-costs (e.g., social tensions) to be negligible and all alternatives far less likely to succeed, it is clear that this line of action complies with strategic rationality. However, it does so at the expense of inferential validity, since a stronger argument (the one based on long-term economical benefits) is silenced, and also with some detriment to dialectical appropriateness, since creating an *ad hoc* polemical target to rally support against it hardly conforms with the procedural rules of civilized debate. So, to put it simply, here we have a case where strategic rationality is satisfied by sacrificing inferential validity and dialectical appropriateness. The question is whether this is normatively legitimate. In other words, is this argumentative strategy rational *overall*, or not?

In this case, the answer is in the affirmative. And what makes all the difference is the context. By "context" here I do not mean anything mysterious, but rather refer to the *ultimate goal* for which a certain argumentative process comes into existence – and such ultimate goal is, typically, extra-dialogical, that is, we argue as a means to achieve something that goes beyond mere "victory" of the argument itself (for details on this point, see Paglieri & Castelfranchi, 2010a).[7] In the example above, and in political discourse in general, the ultimate goal is bringing about policies that are in the common interest. Since that goal is optimally satisfied by the argumentative strategy described before, that strategy is eminently rational, in spite of its shortcomings in terms of inferential validity and dialectical appropriateness. Put it differently, it is the ultimate goal of the argumentative engagement that determines the relative priority of different (and potentially conflicting) normative concerns: in the case of political debate, strategic rationality often trumps other concerns, and rightly so, as long as the public interest is served.

For the same reason, different contexts might instead assign greater weight to other concerns. Imagine for instance that a similar argumentative decision is faced by a teacher aiming to explain students why the reform policies of a certain administration were to be applauded. Here the question of what argument is easier to follow and more likely to garner consensus is nearly irrelevant, whereas extreme care has to be put in presenting valid arguments in support of the teacher's position. Again, what dictates the order of priority among normative concerns is the ultimate goal of the activity: in academic matter, getting to the truth of the matter is often the gold standard, and thus inferential validity trumps other rationality criteria.

This dovetails nicely with the recent application of virtue theory to argumentation: in parallel with the resurgence of interest for virtue ethics and virtue epistemology, it has been proposed (e.g., Cohen, 2005, 2007, 2009; Aberdein, 2007, 2010; Battaly, 2010; for a partial critique, see Bowell & Kingsbury, 2013) that also

---

[7] Besides ultimate goals, another element that plays a key role in defining different argumentative contexts are the *values* that a certain community considers central to that particular situation. For the sake of brevity, here in-depth analysis of the role of values in argumentation is skipped (but see Gilbert, 1997, for an illuminating discussion of this point, and Lloyd Bitzer's work on what he called "the rhetorical situation", 1968, 1980).

argumentation studies should pay greater attention to the character of arguers and the virtues (or vices) that they might express while arguing. Whether argumentative virtues should be considered universal or not is a thorny issue, and no consensus has emerged on a final list of such virtues. But the fact that the *application* of these virtues is context-dependent is hardly disputed: indeed, Cohen's characterization of a sense of proportion as a limiting factor (a meta-virtue?) on the argumentative virtue of open-mindedness points in that very same direction. The implications of his view are more far-reaching, though, since «we need to admit that mere knowledge in and of itself does not really have all that much intrinsic value. What may not be as obvious is that in our recourse to values in explaining this epistemic phenomenon, we are implicitly acknowledging that the fundamental epistemic project –the pursuit of knowledge– needs to be balanced against the rest of our cognitive projects and our other life-projects. A sense of proportion about our epistemic projects will often cut against that particular epistemic grain. It can actually prevent us from acquiring more knowledge» (Cohen, 2009, p. 61). Applied to the present discussion, this suggests that the priority given to strategic considerations in our policy-making example is based on the underlying virtues expressed by that particular strategy: more generally, it is based on ethical values, whether or not one favors virtue theory in ethics and argumentation. In that example, strategic rationality trumps inferential validity and dialectical appropriateness not because it serves the personal interests of the politicians in question, but because it best satisfies the value politicians are supposed to uphold as their most sacred goal, i.e. the common interest. In that context, it is virtuous to relax one's obligations towards validity and appropriateness, while it would be vicious to sacrifice a valuable policy to such scruples, no matter how epistemologically important they are.

We can now appreciate why strategic rationality is not necessarily subordinated to inferential validity and dialectical appropriateness, when it comes to assessing the overall normative legitimacy of argumentative decisions. This also implies that such decisions deserve proper recognition in argumentation theories, not only as a matter of empirical curiosity, but also as a source of interesting normative intuitions. Hopefully, no fellow scholar shall lose sleep over that prospect, since what is being advocated is just a broadening of the field in general, with no specific obligations placed on any individual. On the contrary, I expect logicians with an interest for argumentation to maintain their privileged focus on inferential validity, whereas colleagues working in pragmatics will mostly keep on studying dialectical rules and practices. However, they will have to be joined by a relatively new group of "psychologists of argumentation", with a keen interest also (not only) for argumentative decisions. They should be welcomed in the broad family of argumentation scholars, not just as mere collectors of facts, but as argumentation theorists in their own right, since their paramount aim will be to outline a psychological theory of argumentative processes.[8]

---

[8] For this to happen, psychologists will have to ally with other scholars interested in the psychological effects of communicative acts: marketing and advertising are obvious areas of relevance here, but also

5. Conclusion

In this paper I highlighted the relative neglect for the study of argumentative decisions in argumentation theories, I insisted in contrast on its potential usefulness, summarizing some recent results in that direction, and I defended the decision-theoretic approach to argumentation against the objection that it lacks in normative ambitions.

Admittedly, these are just few preliminary steps in a far-reaching and (hopefully) fruitful research programme. But why should we venture along that uncertain path, instead of treading more familiar and well-respected avenues? I see two main reasons for taking this scientific bet. On the one hand, the study of argumentative decisions naturally lead to integrating theoretical analysis and empirical research, as we have seen even in this short contribution: such integration is both necessary and productive, as others have convincingly argued (Hample, 2005) and/or practically demonstrated (e.g., see the recent empirical study on argument reinstatement, published in *Cognitive Science*; Rahwan et al. 2010). On the other hand, argumentative decisions are crucial for various practical applications of argumentation theories, ranging from the development of new argument-based technologies (for discussion, see Paglieri & Castelfranchi, 2010a) to educational reforms, e.g. for developing critical thinking skills and curricula (for a survey, see Sobocan & Groarke, 2009).

Upon reflection, the reason why people keen to improve their argumentative competence turn to the self-help aisle of their favourite book store (featuring masterpieces such as *How to argue like Jesus*), instead of listening to the wisdom of argumentation theories, is because the latter so far failed to explain, or even try to, how arguers can select optimal moves in real-life debates. One does not need to wish for mass conversion of sedated argumentation scholars into hardened spin doctors, to appreciate that such lack of interest is unwarranted and ultimately disastrous. The point is not to produce yet another list of dialectical tricks to "win every time" or "argue powerfully, persuasively, positively" (as per the subtitles of two successful self-help manuals): such pamphlets have been in circulation since antiquity, from Quintilian's *Institutio Oratoria* to Schopenauer's *The Art of Always Being Right*, without improving too much our understanding of argumentative decisions. The real challenge is to explain *why* such tricks work: how, when, and under what circumstances. Argumentation theories are ideally equipped to tackle this challenge, but they need first to pay closer attention to argumentative decisions and their underlying rationality.

---

communication studies in general. Maurice Finocchiaro recently emphasized the same point, while commenting on the current disregard for experimental studies in argumentation theories: «although a philosopher would want to adopt a critical stance toward such empirical work, it is clearly suggestive and one can ignore it only at one's own risk» (2011, p. 252).

Acknowledgements

This research was supported by the project *PRISMA - PiattafoRme cloud Interoperabili per SMArt-government*, funded by the Italian Ministry of Education, University and Research (MIUR), and by the research network *SINTELNET - European Network for Social Intelligence*, funded by the European Union. I am indebted to Cristiano Castelfranchi, Dale Hample, Steven Patterson, Jean Goodwin, Maurice Finocchiaro, John Woods, Steve Oswald, and Marcin Lewinski for helpful discussion on some of the ideas presented in this paper. A preliminary version of this text was presented at the conference "Virtues of Argumentation" of the Ontario Society for the Study of Argumentation, Windsor, Canada, May 22-25, 2013, and it is available in the electronic proceedings of that event, edited by Dima Mohammed and Marcin Lewinski.

References

Aberdein, A. (2007). Virtue argumentation. In F. H. van Eemeren, A. Blair, C. Willard & B. Garssen (eds.), *Proceedings of ISSA 2006* (pp. 15-20). Amsterdam: SicSat.
Aberdein, A. (2010). Virtue in argument. *Argumentation, 24(2)*, 165-179.
Amgoud, L., & Maudet, N. (2002). Strategical considerations for argumentative agents. In S. Benferhat & E. Giunchiglia (Eds.), *Proceedings of NMR-2002* (pp. 399-407). Toulouse: IRIT.
Battaly, H. (2010). Attacking character: Ad hominem argument and virtue epistemology. *Informal Logic, 30*, 361-390.
Benoit, P., & Benoit, W. (1990). To argue or not to argue: How real people get in and out of interpersonal arguments. In R. Trapp & J. Schuetz (Eds.), *Perspectives on Argumentation: Essays in Honor of Wayne Brockriede* (pp. 55-72). Prospect Heights: Waveland.
Benoit, W., & Benoit, P. (1987). Everyday argument practices of naïve social actors. In J. Wenzel (Ed.), *Argument and Critical Practices* (pp. 465-474). Annandale: Speech Communication Association.
Benthem, J. van (2009). One logician's perspective on argumentation. *Cogency, 1(2)*, 13-25.
Bitzer, L. (1968). The rhetorical situation. *Philosophy and Rhetoric, 1*, 1-14.
Bitzer, L. (1980). Functional communication: A situational perspective. In E. E. White (Ed.), *Rhetoric in transition: Studies in the nature and uses of rhetoric* (pp. 21-38). University Park: Pennsylvania State University Press.
Bowell, T., & Kingsbury, J. (2013). Virtue and argument: Taking character into account. *Informal Logic, 33(1)*, 22-32.

Cionea, I., Hample, D., & Paglieri, F. (2011). A test of the argument engagement model in Romania. In F. Zenker (Ed.), *Argumentation: Cognition & Community. Proceedings of OSSA 2011*. CD-ROM, Windsor: OSSA.

Cohen, D. (2005). Arguments that backfire. In D. Hitchcock (Ed.), *The Uses of Argument. Proceedings of OSSA 2005* (pp. 58-65). Hamilton: OSSA.

Cohen, D. (2007). Virtue epistemology and argumentation theory. In H. Hansen, C. Tindale, R. Johnson & J.A. Blair (Eds.), *Dissensus and the Search for Common Ground*. CD-ROM, Windsor: OSSA.

Cohen, D. (2009). Keeping an open mind and having a sense of proportion as virtues in argumentation. *Cogency, 1(2)*, 49-64.

Eemeren, F. van, & Grootendorst, R. (2004). *A Systematic Theory of Argumentation: The Pragma-Dialectical Approach*. Cambridge: Cambridge University Press.

Eemeren, F. van, & Houtlosser, P. (2002). Strategic maneuvering: Maintaining a delicate balance. In: F. van Eemeren & P. Houtlosser (Eds.). *Dialectic and Rhetoric: The Warp and Woof of Argumentation Analysis* (pp. 131-159). Dordrecht: Kluwer.

Eemeren, F. van, & Houtlosser, P. (2006). Strategic maneuvering: A synthetic recapitulation. *Argumentation, 20*, 381-392.

Finocchiaro, M. (2011). Conductive arguments: A meta-argumentation approach. In J.A. Blair & R. Johnson (Eds.), *Conductive argument. An overlooked type of defeasible reasoning* (pp. 224-261). London: College Publications.

Gabbay, D., & Woods, J. (2003). Normative models of rational agency. *Logic Journal of the IGPL, 11*, 597-613.

Gilbert, M. (1997). *Coalescent Argumentation*. Mahwah: Lawrence Erlbaum Associates.

Ginet, C. (2001). Deciding to believe. In M. Steup (Ed.), *Knowledge, Truth, and Duty* (pp. 63-76). Oxford: Oxford University Press.

Goodwin, J. (2005). What does arguing look like? *Informal Logic, 25*, 79-93.

Goodwin, J. (2007). Argument has no function. *Informal Logic, 27*, 69-90.

Habermas, J. (1984). *The Theory of Communicative Action*. Boston: Beacon Press.

Hample, D. (2005). *Arguing: Exchanging Reasons Face to Face*. Mahwah: Lawrence Erlbaum Associates.

Hample, D., & Benoit, P. (1999). Must arguments be explicit and violent? A study of naive social actors' understandings. In F. van Eemeren, R. Grootendorst, J.A. Blair & C. Willard (Eds.), *Proceedings of ISSA 1998* (pp. 306-310). Amsterdam: SicSat.

Hample, D., & Dallinger, J. (1990). Arguers as editors. *Argumentation, 4*, 153-169.

Hample, D., & Dallinger, J. (1992). The use of multiple goals in cognitive editing of arguments. *Argumentation and Advocacy, 28*, 109-122.

Hample, D., Benoit, P., Houston, J., Purifoy, G., Vanhyfte, V., & Wardwell, C. (1999). Naive theories of argument: Avoiding interpersonal arguments or cutting them short. *Argumentation and Advocacy, 35*, 130-139.

Hample, D., Paglieri, F., & Na, L. (2011). The costs and benefits of arguing: Predicting the decision whether to engage or not. In F. van Eemeren, B. Garssen,

D. Godden & G. Mitchell (Eds.), *Proceedings of ISSA 2010* (pp. 718-732). Sic Sat: Amsterdam.

Hample, D., Werber, B., & Young, D. (2009). Framing and editing interpersonal arguments. *Argumentation, 23*, 21-37.

Hicks, D. (1991). A descriptive account of interpersonal argument. In D.W. Parson (Ed.), *Argument in Controversy* (pp. 167-181). Annandale: Speech Communication Association.

Infante, D., & Rancer, A. (1982). A conceptualization and measure of argumentativeness. *Journal of Personality Assessment, 46*, 72-80.

Infante, D., & Wigley, C. (1986). Verbal aggressiveness: an interpersonal model and measure. *Communication Monographs, 53*, 61-69.

Johnson, R. (2000). *Manifest Rationality: A Pragmatic Theory of Argument*. Mahwah: Lawrence Erlbaum Associates.

Karunatillake, N., & Jennings, N. (2005). Is it worth arguing? In I. Rahwan, P. Moratïs & C. Reed (Eds.), *Argumentation in Multi-Agent Systems. Proceedings of ArgMAS'04* (pp. 234-250). Berlin, Springer.

Martin, R.W., & Scheerhorn, D.R. (1985). What are conversational arguments? Toward a natural language user's perspective. In J.R. Cox, M.O. Sillars & G.B. Walker (Eds.), *Argument and social practice* (pp. 705-722). Annandale: Speech Communication Association.

O'Keefe, D. (1977). Two concepts of argument. *Journal of the American Forensic Society, 13*, 121-128.

Paglieri, F. (2007). No more charity, please! Enthymematic parsimony and the pitfall of benevolence. In H. Hansen, C. Tindale, R. Johnson & J.A. Blair (Eds.), *Dissensus and the Search for Common Ground* (pp. 1-26). CD-ROM, Windsor: OSSA.

Paglieri, F. (2009). Ruinous arguments: Escalation of disagreement and the dangers of arguing. In J. Ritola (Ed.), *Argument Cultures: Proceedings of OSSA 2009* (pp. 1-15). CD-ROM, Windsor: OSSA.

Paglieri, F. (2010). Committed to argue: On the cognitive roots of dialogical commitments. In C. Reed & C. Tindale (Eds.), *Dialectics, Dialogue and Argumentation. An Examination of Douglas Walton's Theories of Reasoning* (pp. 59-71). London: College Publications.

Paglieri, F. (2013). Choosing to argue: Towards a theory of argumentative decisions. *Journal of Pragmatics*, 59(B), 153-163.

Paglieri, F., & Castelfranchi, C. (2005). Influence of social motivation over belief dynamics: A game-theoretical analysis. In B. Kokinov (Ed.), *Advances in Cognitive Economics* (pp. 202-213). Sofia: NBU Press.

Paglieri, F., & Castelfranchi, C. (2010a). Why argue? Towards a cost-benefit analysis of argumentation. *Argument & Computation, 1*, 71-91.

Paglieri, F., & Castelfranchi, C. (2010b). In parsimony we trust: Non-cooperative roots of linguistic cooperation. In A. Capone (Ed.), *Perspectives on Language Use and Pragmatics* (pp. 99-117). München: Lincom Europa.

Paglieri, F., & Woods, J (2011a). Enthymematic parsimony. *Synthese, 178*, 461-501.

Paglieri, F., & Woods, J (2011b). Enthymemes: From reconstruction to understanding. *Argumentation, 25(2)*, 127-139.

Rahwan, I., & Larson, K. (2009). Argumentation and game theory. In I. Rahwan & G. Simari (Eds.), *Argumentation in Artificial Intelligence* (pp. 321-339). Berlin: Springer.

Rahwan, I., Madakkatel, M., Bonnefon, J.-F., Awan, R., & Abdallah, S. (2010). Behavioural experiments for assessing the abstract argumentation semantics of reinstatement. *Cognitive Science, 34(8)*, 1483-1502.

Reed, C., & Walton, D. (2003). Argumentation schemes in argument-as-process and argument-as-product. In J.A. Blair, D. Farr, H. Hansen, R. Johnson & C. Tindale (Eds.), *Informal Logic @ 25: Proceedings of OSSA 2003*. Windsor: OSSA.

Riveret, R., Prakken, H., Rotolo, A., & Sartor, G. (2008). Heuristics in argumentation: A game-theoretical investigation. In P. Besnard, S. Doutre & A. Hunter (Eds.), *Proceedings of COMMA 2008* (pp. 324-335). Amsterdam, IOS Press.

Ryan, S. (2003). Doxastic compatibilism and the ethics of belief. *Philosophical Studies, 114*, 47-79.

Sobocan, J., & Groarke, L. (Eds.) (2009). *Critical Thinking Education and Assessment: Can Higher Order Thinking Be Tested?* London, ON: Althouse Press.

Sperber, D., & Wilson, D. (1986). *Relevance: Communication and Cognition.* Oxford: Blackwell.

Steup, M. (2000). Epistemic deontology and the voluntariness of belief. *Acta Analytica, 15*, 25-56.

Toulmin, S.E. (1958). *The Uses of Argument*. Cambridge: Cambridge University Press.

Vuchinich, S. (1990). The sequential organization of closing in verbal family conflict. In A.D. Grimshaw (Ed.), *Conflict Talk* (pp. 118-138). Cambridge: Cambridge University Press.

Walton, D. (1996). *Argumentation Schemes for Presumptive Reasoning*. Mahwah: Lawrence Erlbaum Associates.

Walton, D., & Godden, D. (2007). Informal logic and the dialectical approach to argument. In H. Hansen & R. Pinto (Eds.), *Reason Reclaimed. Essays in honor of J. Anthony Blair and Ralph H. Johnson* (pp. 3-17). Newport News: Vale Press.

Walton, D., & Krabbe, E. (1995). *Commitment in Dialogue: Basic Concepts of Interpersonal Reasoning*. Albany: SUNY Press.

Wansing, H. (2000). A reduction of doxastic logic to action logic. *Erkenntnis, 53*, 267-283.

Wansing, H. (2002). Seeing to it that an agent forms a belief. *Logic and Logical Philosophy, 10*, 185-197.

Wansing, H. (2006). Doxastic decisions, epistemic justification, and the logic of agency. *Philosophical Studies, 128*, 201-227.

Wilson, D., & Sperber, D. (2002). Truthfulness and relevance. *Mind, 111*, 583-632.

Woods, J. (2005). Epistemic bubbles. In S. Artemov, H. Barringer, A. d'Avila Garcez, L. Lamb & J. Woods (Eds.), *We Will Show Them: Essays in Honour of Dov Gabbay* (vol. II, pp. 731-774). London: College Publications.

Part II
Arguing on the Web

Chapter 4

Arguers and the Argument Web

Mark Snaith, Rolando Medellin, John Lawrence, & Chris Reed

School of Computing, University of Dundee, DUNDEE, DD1 4HN, UK
{marksnaith, rmedellin, johnlawrence,chris}@computing.dundee.ac.uk

Abstract. In this paper, we lay some groundwork in understanding the ways in which users of the Argument Web can be represented in the Argument Interchange Format (AIF). In particular, we tackle two specific challenges of (i) how to handle various roles on the Argument Web, based on the level and type of contribution they make; and (ii) how to support mixed initiative argumentation in which software agents may adopt a role defined by the previous activity of human arguers. We then go on to present an extra layer of the Argument Web, the Social Layer, which handles both the representation of users and connections to the social World Wide Web.

1. Introduction

The relationships between arguers and the arguments that they create, navigate and invoke are many and complex. As the number of tools available on the Argument Web continues to increase, as the datasets available there expand, and as the user base grows, these relationships need to be understood and defined so that software can work with them coherently and consistently. Here, we lay some groundwork in understanding the ways in which users need to be represented, particularly in order to tackle two specific challenges: (i) how to handle various roles on the Argument Web, based on the level and type of contribution they make; and (ii) how to support mixed initiative argumentation in which software agents may adopt a role defined by the previous activity of human arguers.

The paper proceeds as follows: in section 2, we provide brief introductions to the Argument Web and the Argument Interchange Format; in section 3, we characterize three different roles on the Argument Web; in section 4 we specify our method for representing users in the AIF; in section 6 we show how representing roles can be deployed in a practical application and in section 7 we conclude the paper.

2. Background

2.1. Argument Web

The Argument Web is a vision for integrated, reusable, semantically rich resources connecting views, opinions, arguments and debates online, wherever they may occur, whether in blogs, in comments or in multimedia resources, whether in educational, political, legal or other domains (Rahwan, 2008; Rahwan, Zablith, & Reed, 2007; Bex et al., 2013).

Recent work has seen development of Argument Web infrastructure (Lawrence et al., 2012) and applications (Snaith et al., 2012; Lawrence, Bex, & Reed, 2012), and the platform is expected to grow rapidly in the near future.

2.2. Argument Interchange Format

The Argument Web is underpinned by the Argument Interchange Format (AIF) which is a description, standard and series of implementations of a mechanism for exchanging argument resources between tools (Chesñevar et al., 2006). The AIF is specified by a core ontology, which is implemented in various forms. One such form is a MySQL[14] database (AIFdb)[15], the entity-relationship diagram for which can be seen in Figure 1 and illustrates the nature of the ontology.

The main principles of the AIF that are relevant to the present work are **Locutions** and **People**. Locutions are statements (written or oral) that have a property indicating the utterer and are either directly contributed (via an application such as ArguBlogging, see Snaith et al., 2012) or indirectly, through an analysis of what a person has said (via an application such as OVA, see Snaith, Lawrence, & Reed, 2010). Representing locutions in the AIF allows of a degree of attribution, which supports mixed initiative dialogue applications such as Arvina (Lawrence, Bex, & Reed, 2012). However, this representation goes no further than to provide a name; what is not account for is exactly who the utterers of the locutions are. While this is entirely reasonable within the core AIF, since it is a format for representing *argument*, there are advantages to allowing a closer connection between locutions and the people who made them. From the application perspective, it allows systems such as Arvina to harvest all of a user's opinions as opposed to just those available based on the current dialogue; from the user perspective, maintaining a close connection to what they have contributed is behavior that is already established on the conventional Web, and so would be expected of the Argument Web as well. These will be further elaborated in section 4.

---

[14] While MySQL has been used for AIFdb, in principle any SQL-based database management system can be used.
[15] http://aifdb.org

2.3. Mixed initiative argumentation

Mixed initiative argumentation allows both real and virtual (software) agents to participate in a dialogue, with the discussion being driven by any participant.

Arvina is an application that employs mixed initiative argumentation to allow for ease of navigation through complex debates (Lawrence, Bex, & Reed, 2012). The system spawns virtual participants representing the views of people whose arguments are already available on the Argument Web. Human users direct the discussion, asking questions of other participants (whether real or virtual) and can at any stage offer their own opinions on the current topic.

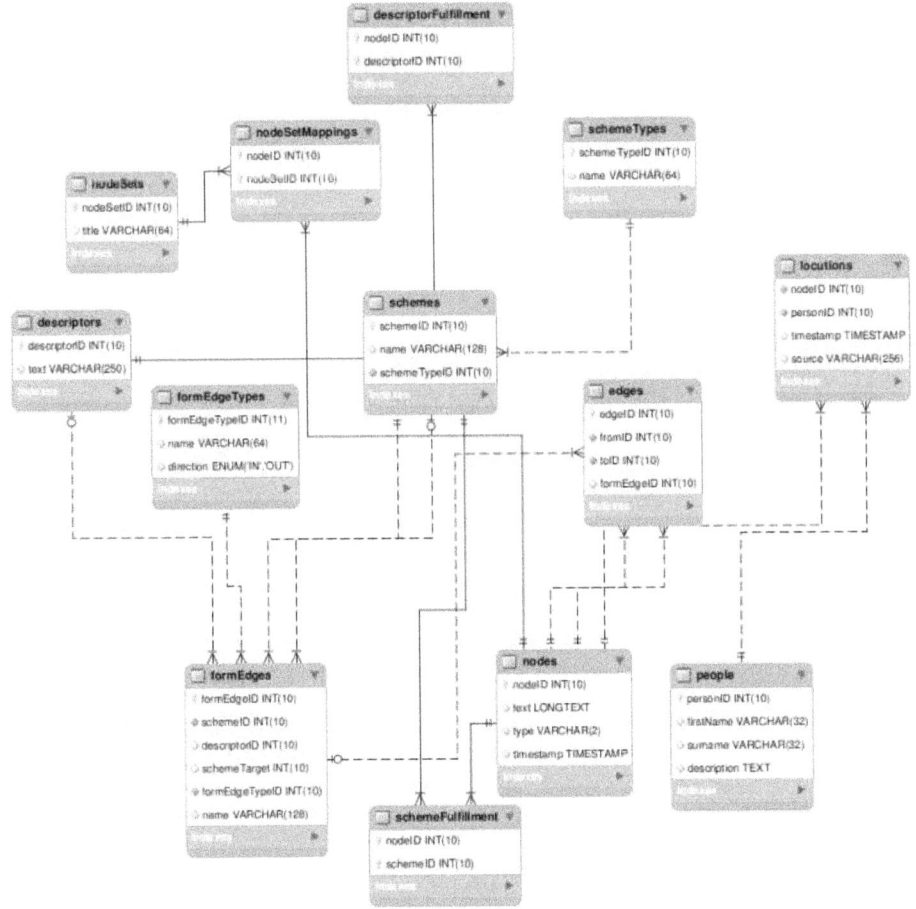

Figure 1. AIF core ontology

3. Roles on the Argument Web

3.1. Speakers

A *speaker* is the person from which the content of a locution was obtained. A speaker is not necessarily a registered user of the Argument Web (see section 3.2), but can instead be an agent (see section 3.3) or the source of a piece of analyzed text. When a connection is made between a speaker and a locution, applications such as Arvina (Lawrence, Bex, & Reed, 2012) are able to spawn virtual agents to represent speakers in a new dialogue (see "Agents" below).

Where a speaker is a registered Argument Web user, there are obvious advantages to maintaining a connection between the user and the locutions they have made. If the speaker (user) were to have an agent spawned from them via Arvina, that virtual representation can link back to their user profile, and subsequently their other contributions to the Argument Web. This in turn allows the spawned agent to access other contributions from that user outside of the nodeset being used to initiate the dialogue.

3.2. Users

A *user* is a person or group that has registered with the Argument Web and contributes through the use of applications. Registration can either be direct, or via a social networking platform such as Facebook or Twitter. A key feature of users is that their contributions can be traced back to them.

Certain applications deployed on the Argument Web already permit user registration, such as ArguBlogging (Snaith et al., 2012), however this is done simply to allow for easy connection to blog and social networking sites and does not maintain a close connection between the users and the AIF resources they create. The only connection is that the user is represented as the speaker of the locution created by contributing to the Argument Web, however this is little more than a text-based property of an AIF L-Node (see section 3.1) and as such there is no distinction between a contribution from, for instance, John Smith in Dundee and John Smith in London.

3.3. Agents

A "Dialogue-Agent" is a representation of *users* and *speakers* in mixed initiative dialogues.

More specifically, "Dialogue-Agents" are entities that represent what *speakers* (from the AIFdb) and *users* (from the Social Layer) said in the context of a debate in the AIFdb. For a given debate or analysis in the AIFdb, we extract the ID's of the speakers that issued locutions and for each speaker, we create a separate entity (the agent) that contains the related assertions and the context in which they were issued. In this way, we can differentiate the actual user or speaker from the agent that represents him in mixed initiative dialogue environments.

In more detail, each agent represents a unique speaker (personID) or user (userID) in a debate or analysis (a nodeset) identified by all its related locution nodes (L-nodes).

We also use the term "Dialogue-Agents" referring to *autonomous agents* in multiagent systems, where an agent is defined as a software entity able to react autonomously employing a representation of its beliefs. The "representation of beliefs" in our case is given only by what speakers or users *said* in the context of a debate or analysis. And the agent "reactions" are extracted from the assertions related to the speaker. In this way, if the agent is questioned, the response can be taken from the assertions related to the user or speaker in the AIFdb node structure (the mechanism to select a particular locution is out of the scope of this paper).

With this mechanism, "Dialogue-Agents" can represent speakers and users from the "Social Layer" in mixed initiative dialogue platforms keeping separate the original analysis in the AIFdb from the new dialogues generated by the agents in other platforms.

Take the example:

- Melanie Philips (MP) says: P is the case
- Michael Buerk (MB) says: Why P?
- MP says: P because of Q
- MB says: Q is not the case therefore P is not the case.
- Claire Fox (CF) says: The evidence of Q is S, so P and Q are the case.
- MB says: I accept that.
- MP says: I don't!

From this dialogue we can extract three agents (represented as speakers in the the AIFdb structure): one agent representing Melanie Philips (*MP agent*), another agent representing Michael Buerk (*MB agent*) and an agent representing Claire Fox (*CF agent*). These agents then can be deployed with the information they believe based on their assertions.

MP agent believes P and Q, *MB agent* believes S and Q and *CF agent* believes S. If the Melanie Philips agent is asked in Arvina: Do you agree with Q? The agent can confirm this is the case because there exist a direct assertion in the dialogue but the agent cannot agree with S. Furthermore if the *CF agent* is challenged with: not Q, he can answer with S.

3.4. Arguers

Arguers are the participants in a dialogue that is being executed in an application such as Arvina (Lawrence, Bex, & Reed, 2012). An arguer is either a (human) user of the Argument Web, or a (virtual) dialogue-agent that can respond to human or other agents' locutions. The dialogue execution platform sees all arguers as equal insofar as they all, whether real or virtual, can initiate the dialogue, advance locutions and respond to other participants. This process, called mixed initiative argumentation (Snaith, Lawrence, & Reed, 2010), allows real users to debate with dialogue-agents representing real speakers.

4. Representing Users in the AIF

Representations of Argument Web roles in the AIF is currently limited to a property of locutions that indicate the utterer and to explicitly account for the other roles in the core ontology would be peripheral to the AIF's purpose of representing argument. Nevertheless, representing roles, and users in particular, provides certain advantages. First, it maintains a formal link between AIF resources and a profile representing the person that created them, providing for greater integration between the social World Wide Web and the Argument Web; second, it allows user information to be transmitted as part of the AIF that is used for communication between the Argument Web's applications and underlying infrastructure.

While modifications to the core AIF ontology are undesirable, it remains possible to extend the AIF through the use of adjunct ontologies (AOs). An AO is a specification of application-specific concepts and features that does not change the core AIF, but instead references it. Applications that are aware of the AO call the AO, which then indexes core AIF structures. The main advantage of an AO is that the core AIF does not have any knowledge of the extensions it provides, and as such tools and infrastructure that are capable of processing only core AIF can continue to process AIF extended by the AO, and simply ignore the extensions. We therefore propose an AO for representing roles in the AIF.

In the remainder of this section, we first specify our AO for representing users on the Argument Web, before describing ways of detecting users that have been identified as the speaker of a locution, but did not contribute themselves.

4.1. Social AIF

To represent users in the AIF, we extend the core specification using an adjunct ontology. We describe AIF extended in this way as "Social AIF" or S-AIF, which consists of four concepts:
- **Users**: A user, as described in section 3.2. Firstname and surname are mandatory properties of users, however other information (e.g.

date of birth) may be application-specific. As such, we place no constraints on what information can be stored and delegate responsibility for representing this to UserInfo.
- **UserInfo**: Information about the user, such as date of birth, address, occupation etc. Each instance of UserInfo stores one piece of information. Representing user details in this way places no assumptions or constraints on the level of user information an application may require.
- **Applications**: Applications used by users to create locutions. A user is connected to locutions via an application.
- **Locution mapping**: A mapping of users in the AO to locutions in one or many AIFdb instances, with a record of the application used to create the locutions. A locution ID is a URI pointer to the relevant L-Node in an AIFdb instance.

Similar to the core AIF ontology, the S-AIF AO can be implemented as a MySQL database, whose entity-relationship diagram is shown in Figure 2.

A rendering of a S-AIF diagram, showing how users and applications connect to locutions, is provided in Figure 3.

5. Argument Web Social Layer

The nature of Social AIF as an adjunct ontology means that it is neither interpreted nor stored by AIFdb. Thus in order for S-AIF to be of use, we require a mechanism of storing and retrieving the extensions provided by the AO. To this end, we add a *Social Layer* to the Argument Web. The Social Layer sits between applications and AIFdb, acting as a platform for social interaction on the Argument Web, managing users and maintaining connections to the social World Wide Web, through sites such as Blogger[16], Tumblr[17] and Facebook[18].

The Social Layer both returns and consumes S-AIF. To retrieve S-AIF, a user or application provides a nodeset ID via a RESTful web service. This ID is then passed down into AIFdb to retrieve core AIF; the Social Layer then consults the S-AIF database to append the necessary user and application information, converting the AIF into S-AIF.

Consumption of S-AIF involves posting it in JSON format to a RESTful web service. The core AIF is then separated from the S-AIF extensions and stored in AIFdb, with the mapping of users to locutions and applications being stored in the S-AIF database. Additionally, if S-AIF that is posted to the Social Layer is to also be published to a social media platform (Blogger, Facebook, Tumblr etc.), it is passed to a translation module that converts the JSON structure into a textual summary of

---

[16] http://www.blogger.com
[17] http://www.tumblr.com
[18] http://www.facebook.com

the argument, appropriate to both the platform it is being sent to, and the application that generated the S-AIF.

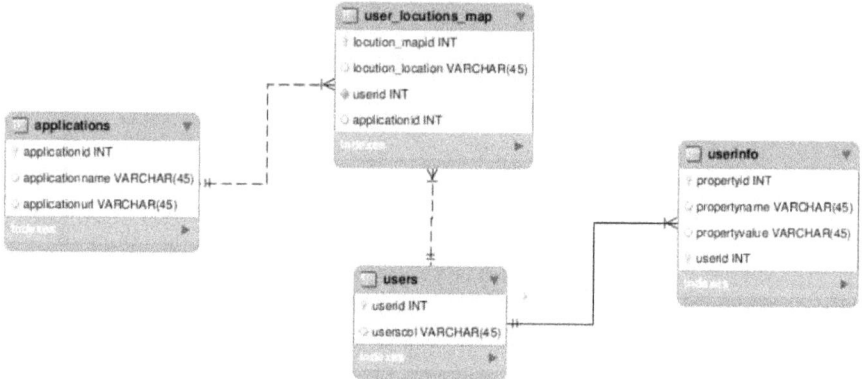

Figure 2. S-AIF adjunct ontology

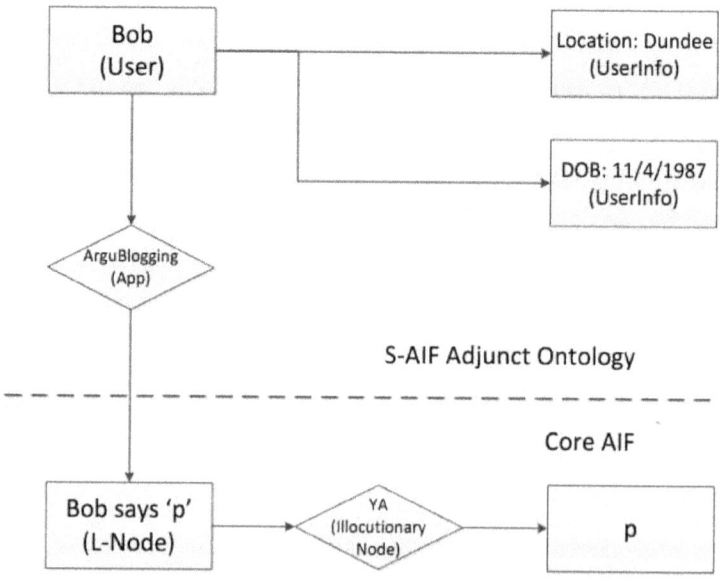

Figure 3. S-AIF connections to core AIF

The Social Layer architecture has been designed such that it can be extended with new social media platforms, along with appropriate formatting algorithms. The architecture of the Social Layer is shown in figure 4.

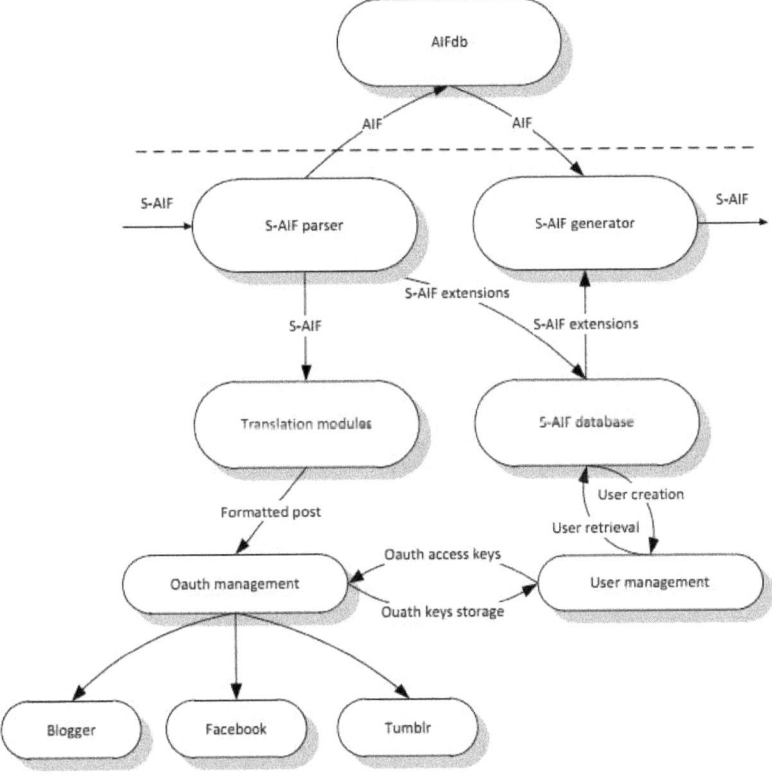

Figure 4. Argument Web Social Layer Architecture

6. Practical Applications

In this section, we give examples of two practical applications of S-AIF and the Argument Web Social Layer, one that uses S-AIF and the Argument Web Social Layer to maintain the connection between users and their contributions (ArguBlogging), and another that uses S-AIF resources to enhance the experience of mixed-initiative dialogues (Arvina).

6.1. ArguBlogging

ArguBlogging is a tool (shown in Figure 5 that allows users to respond to online articles in a structured way, with their response being channeled to both their blog and the Argument Web (Snaith et al., 2012). The current version of ArguBlogging generates only core AIF and so the link between the user, their blog and the Argument Web resources is weak at best — the only connection is the URL of the blog post being attached to the AIF locution.

Using S-AIF and the Argument Web Social layer, we have produced a new version of ArguBlogging where these connections are captured. Users of ArguBlogging are now registered users of the Argument Web. When the application is used, it generates S-AIF that is posted to and processed by the Social Layer, resulting in a formal link between the user and their opinion, while also posting to social media platforms.

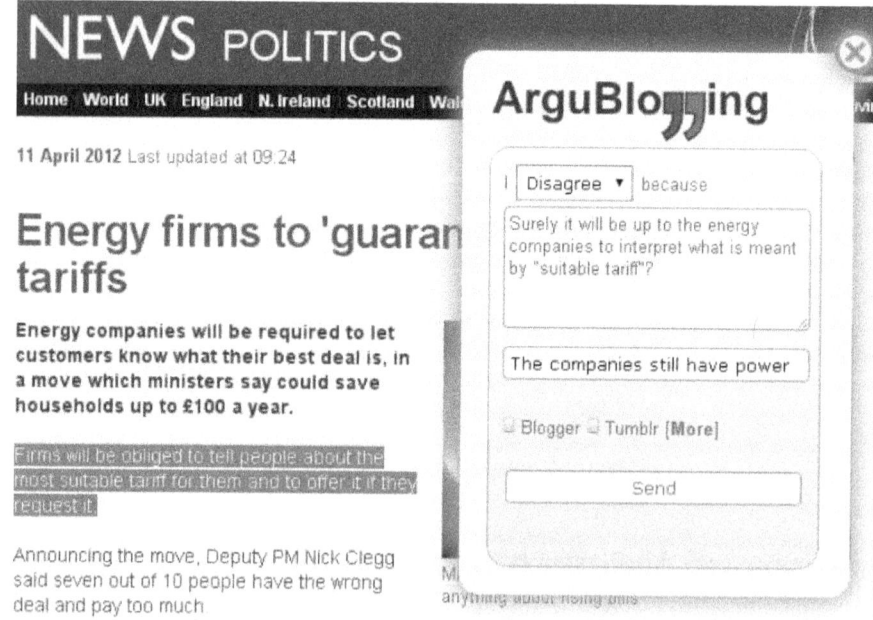

Figure 5. ArguBlogging tool rendered on a web page

A formatting algorithm generates a textual description of the argument that is suitable for posting to a blog; consider as an example the AIF diagram in Figure 6, that was generated from the example shown in Figure 5. The translation module that posts to Tumblr generates the following textual description that is published as a blog post:

"Firms will be obliged to tell people about the most suitable tariff for them and to offer it if they request it"

In the conversation, I disagree because: Surely it will be up to the energy companies to interpret what is meant by "suitable tariff".

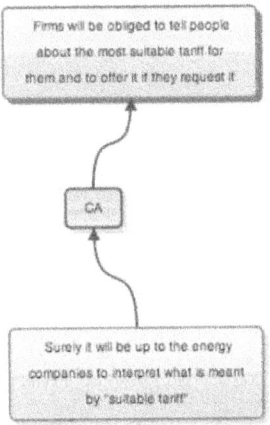

Figure 6. AIF diagram of ArguBlogging argument

6.2. Arvina

Arvina is a software tool for mixed initiative argumentation that provides a dialogical interface to complex debates (Lawrence, Bex, & Reed, 2012). Arvina uses dialogue protocols to structure the discussion between participants, who may either be real users or virtual agents representing the arguments of specific authors who have previously added their opinions to the Argument Web.

An Arvina session is initiated by a user selecting a topic, which represents an I-Node in the Argument Web. Arvina then interacts with AIFdb instances to retrieve an AIF structure containing arguments for and against the selected topic. Using the locutions in this structure, Arvina spawns software agents based on the speakers attached to those locutions. The architecture for this is shown in Figure 7.

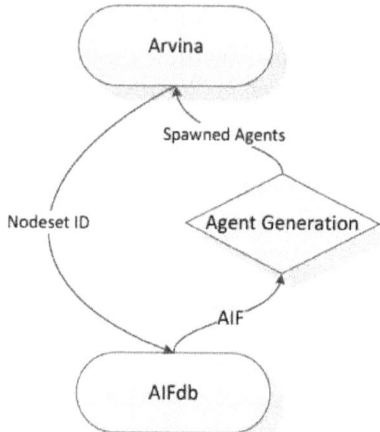

Figure 7. Current Arvina architecture

This architecture has limitations from both the software and user perspectives; the Arvina software is provided with only the name of the speaker of each locution, and is therefore the only information available when spawning agents. For users, there is no way to dig deeper into someone's argument web contributions, because there is no link between what their agent believes in the current dialogue and their other contributions to the Argument Web.

By using the Argument Web Social Layer between Arvina and AIFdb, we can use S-AIF to establish if there exist links between users and the locutions in the nodeset being used to initiate the dialogue. Instead of passing a nodeset ID directly to AIFdb, it is passed to the Social Layer, which then passes it to AIFdb. The Social Layer uses locutions in the returned AIF to query the users database and return S-AIF containing, where applicable, details of any users represented in the nodeset. The S-AIF is passed to the agent generator, which uses the user information to generate the agents, including personal information — for instance, attaching avatars, occupations and locations.

The proposed new architecture is shown in Figure 8.

Providing Arvina with the ability to gather extensive information about the users represented by agents allows for more sophisticated time-shifted dialogues. From an application perspective, Arvina can use the information to harvest all of a user's arguments on the argument web, which in turn expands the knowledge base of the agent representing them. For users of Arvina, they too benefit from the agents possessing a larger knowledge base, but also gain a better insight into the "person" (agent) with whom they are arguing.

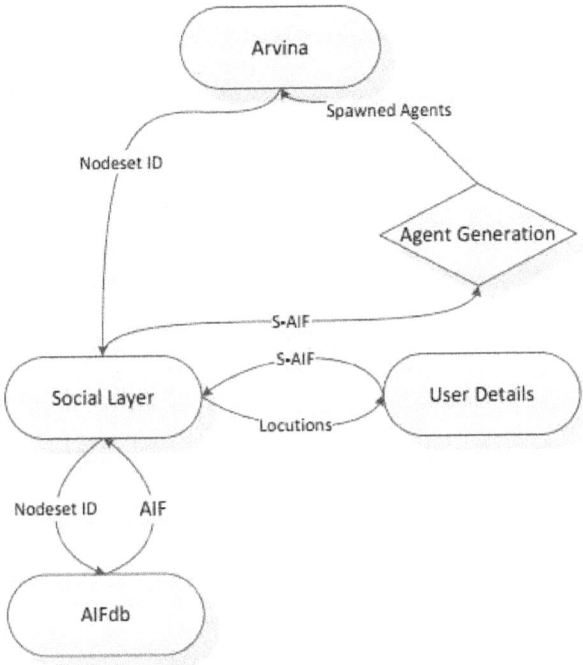

Figure 8. Proposed Arvina architecture using S-AIF

7. Conclusions and Future Work

In this paper, we identified and described three different roles on the argument web — *speakers*, *users* and *agents* — and specified an AIF adjunct ontology (AO) that allows them to be represented on the argument web and connected to the resources they create.

The work in this paper presents only the first steps towards representing roles on the argument web. Future work will focus both on theoretical extensions to the present work, and its practical deployment.

From a theoretical standpoint, we need to consider how users can be connected to argument web resources to which they attributed, but did not contribute — consider, for instance, a user publishing a blog post that is subsequently analysed using an argument mapping tool. While the analyst can identify the user as a speaker, there is no explicit connection between the user and their argument. In addressing this, it is our intention to explore connections to Named Entity Recognition (NER) (Snaith et al., 2012) and Friend-Of-A-Friend (FOAF)[19].

---

[19] http://www.foaf-project.org/

Before the work presented can be deployed on the Argument Web, privacy and data protection issues must be considered. Specifically, users must be given the means to control third-party access to their argument web content, where third parties can include other users, general users of the web and software systems. We also intend to explore connections between the social world wide web and the social argument web, and how sites such as *Facebook* and *Twitter* can be used as platforms to support ubiquitous debate.

Handling identities in the Argument Web is surprisingly challenging – we need to distinguish users from the speakers associated with locutions (because those locutions may have been created in domains not mediated by the Argument Web), and also from the agents that may subsequently represent them in online dialogues. Our goal here has been to sketch why the problem is important and how a coherent solution can be formed. This is a vital step as the Argument Web starts to come out of the lab and find application in the real world.

References

Bex, F., Lawrence, J., Snaith, M., & Reed, C. (2013). Implementing the argument web. Communications of the ACM, 56(10), 66-73.

Chesñevar, C., Modgil, S., Rahwan, I., Reed, C., Simari, G., South, M., ... & Willmott, S. (2006). Towards an argument interchange format. The Knowledge Engineering Review, 21(04), 293-316.

Lawrence, J., Bex, F., & Reed, C. (2012). Dialogues on the Argument Web: Mixed Initiative Argumentation with Arvina. In Verheij, B., Szeider, S., & Woltran, S. (Eds) *Proceedings of the Fourth International Conference on Computational Models of Argument (COMMA 2012)*, pp. 513–514, Vienna, Austria. IOS Press.

Lawrence, J., Bex, F., Reed, C., & Snaith, M. (2012). AIFdb: Infrastructure for the Argument Web. In Verheij, B., Szeider, S., & Woltran, S. (Eds) *Proceedings of the Fourth International Conference on Computational Models of Argument (COMMA 2012)*, pp. 515–516, Vienna, Austria. IOS Press.

Nadeau, D., & Sekine, S. (2007). A survey of named entity recognition and classification. Lingvisticae Investigationes, 30(1), 3-26.

Rahwan, I. (2008). Mass argumentation and the semantic web. Web Semantics: Science, Services and Agents on the World Wide Web, 6(1), 29-37.

Rahwan, I., Zablith, F., & Reed, C. (2007). Laying the foundations for a world wide argument web. Artificial intelligence, 171(10), 897-921.

Snaith, M., Bex, F., Lawrence, J., & Reed, C. (2012). Implementing ArguBlogging. In Verheij, B., Szeider, S., & Woltran, S. (Eds) *Proceedings of the Fourth International Conference on Computational Models of Argument (COMMA 2012)*, pp. 511-512, Vienna, Austria. IOS Press.

Snaith, M., Lawrence, J., & Reed, C. (2010). Mixed initiative argument in public deliberation. Online Deliberation, 2. In De Cindo, F., Macintosh, A., & Peraboni, C. (Eds) *From e-Participation to Online Deliberation: Proceedings of the fourth international conference on Online Deliberation (OD2010)*, Leeds, UK.

Chapter 5

From Argumentation Theory To Social Web Applications

Lucas Carstens, Valentinos Evripidou, & Francesca Toni

Department of Computing, Imperial College London, UK,
{lucas.carstens10;valentinos.evripidou10;ft}@imperial.ac.uk

Abstract. In recent years the Argumentation community has seen a growing body of work that is concerned with the ever important question of how one's research may be put to concrete use. With this paper we reflect on our experience with developing concrete applications that make use of Argumentation to various ends. We discuss developments in social web debating settings, as well as potential uses of Natural Language Processing to support the proliferation of argumentative analysis. Based on our experience with these developments we discuss five challenges that we believe to be crucial to the development of useful and usable Argumentation applications. We are developing a number of applications and for each we discuss whether and how we have addressed these challenges and, if not, how we may do so in the future.

1. Introduction

Debating, as well as the act of putting forth arguments, is an inherently social process. Whether we exchange opinions with someone or we make an argument for or against something, Argumentation is always carried out in a social environment through which it is endowed with meaning (why, for example, would we write a film review if not to convince someone else of the movie's merits or demerits?). This means that, if we are to develop Argumentation applications that cater to *real-life* Argumentation, we often need to consider how to support and/or emulate these social processes. With this paper we present five major challenges to building *useful and usable Social Argumentation Web applications*, as well as how they inter-relate and how we address them. We believe there is a need to (1) enable debates by building interfaces to high standards. Additionally we need to (2) provide rigorous frameworks to analyze and quantify the quality of debates and individual arguments, we must (3) consider ways of automating the construction of Argument Frameworks from existing texts and we need to (4) take into account the source of arguments when evaluating their value and validity. Finally, we must (5) consider ways of incentivizing potential users to take up our applications. These challenges, discussed in detail in section 2, highlight the need for building useful and usable applications, analyzing debates sensibly, identifying arguments in text and paying heed to the characteristics of the individuals participating in debates.

The remainder of this paper is organized as follows. In section 2 we discuss the challenges summarized in figure 1. We then discuss the functionalities of Quaestio-it, a social debating web application, in section 3. We introduce Desmold, an Injection Molding decision support system, and Arg&Dec, a more general purpose decision support system, in section 4, both intended as web applications that support design and decision making in engineering. In section 5 we summarize ways of how Quaestio-it is helping the development of Argumentation Mining tools and how Argumentation Mining may contribute to improving our applications. Before concluding the paper in section 7 we present a Use Case for an E-Learning platform in section 6. With this platform we aim to incorporate the concepts introduced in sections 2 to 5 to develop a more interactive experience for all stakeholders in a learning environment.

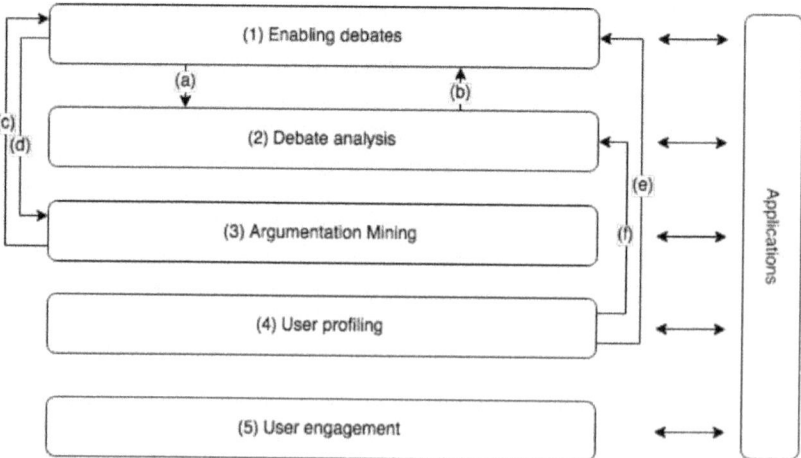

Figure 1. Overview of major challenges to building Online Argumentation applications, as well as links between them.

2. Challenges of Building Applications for Online Debates

Arguing on the web can take a myriad of forms, and with each we face different challenges. Below we describe some of the most central and commonplace problems we need to address to facilitate successful web-based debates.

(1) Enabling debates. First and foremost in the quest to fostering online debates stands the development of applications that build on quality interfaces and back end architecture. To obtain argumentative structures we need to develop applications that guide debating processes in a principled manner. Many formalisms have been proposed to represent arguments, both those created using free natural language and those somewhat more structured. *Argumentation Schemes* (Walton, 2013) have been used to represent arguments in a number of domains, e.g. to facilitate the analysis of legal texts (Wyner & Bench-Capon, 2007; Wyner, Mochales-Palau, Moens, & Milward, 2010). The *Argument Interchange Format (AIF)* (Chesñevar et al., 2006) has been designed to facilitate formal exchange of data for both Argumentation tools and Agent-based systems. On its basis the *AIFdb* (Lawrence, Bex, Reed, & Snaith, 2012) has been developed, a database service that aims at building an *Argument Web*. A second major aspect to enabling debates is the development of interfaces to support debates. We present one such interface in section 3. Similarly to this interface, Cohere (Buckingham Shum, 2008) (cohere.open.ac.uk) and Debatepedia (debatepedia.idebate.org/) amongst others, provide websites that aim at allowing and facilitating debates. Offline applications include Araucaria (Reed & Rowe, 2004) and a tool to support the analysis of arguments in product reviews (Wyner, Schneider, Atkinson, & Bench-Capon, 2012).

(2) Analyzing debates. To add value to debates that goes beyond the traditional architecture of simple discussion fora we need to develop ways of (automatically) analyzing debates. Analysis of debates and/or individual arguments may aim to identify good arguments, baseless arguments, the soundness of chains of arguments, etc. Argumentation research has produced numerous frameworks in which we can analyze debates, including, but not limited to, *Abstract Argumentation (AA)* (Dung, 1995), *Assumption Bases Argumentation (ABA)* (Dung, Kowalski, & Toni, 2009), *Value Based Argumentation (VBA)* (Bench-Capon, 2002), etc. When developing practical applications we need to consider which of those frameworks offers us the best ways of analyzing the arguments we are dealing with. The decision on which framework(s) to use may vary from application to application.

(3) Argumentation Mining. On the one hand we need to be able to analyze debates which are formally marked as such. On the other hand the web contains virtually limitless resources of arguments and debates which are not created with Argumentation research in mind. Customer reviews, editorial opinion pieces and many other texts contain arguments. With Argumentation Mining we aim to develop ways of identifying argumentative structures on the web that are not created in a rigorous debate setting. Argumentation Mining is, most broadly, concerned with identifying arguments and relations that hold between them in natural language text.

Much of the work in this field has focused on the analysis of online content. (Park & Cardie, 2014) identify different classes of argumentative propositions in online user comments, while (Ghosh, Muresan, Wacholder, Aakhus, & Mitsui, 2014) analyze multilogue, instead, classifying relations between user comments. Other application areas for Argumentation Mining include the biomedical (Green, 2014; Houngbo & Mercer, 2014) and the legal domains (Mochales & Moens, 2008; Palau & Moens, 2009; Wyner et al., 2010).

(4) User profiling. An important aspect of fostering debate is the consolidation of arguments from various parties. An example we discuss in section 6 is the exchange between students and lecturers. In such scenarios it is vital to establish which stakeholders partake in a debate and to potentially treat arguments made by parties differently to reflect the status they hold in a debate. To achieve this and to integrate it with the (ideally automatic) analysis of debates we need to establish relevant user profiles that reflect the roles participants have in a debate.

(5) User engagement. An issue that somewhat spans across all applications and development aspects is that of how we may engage individuals to use an application. The success of an application unequivocally depends on whether it is adopted by users. On the one hand this means building quality applications that work. It may, however, not suffice to build an application that works. Often times it is equally important to engage users and incentivize them to both start using an application and to continue using it, subsequently.

As we show in figure 1 the five aspects highlighted here interlink and need to be considered in unison if we are to develop quality Argumentation applications. The creation of debates usually directly determines how we analyze them (a). The more we restrict how users can contribute to a debate and the more additional information we ask of them (e.g. how one's argument relates to other arguments), the easier and more detailed an analysis may be. In turn, the way we want to analyze debates may determine the requirements of debating interfaces (b).

Mining for arguments in text may lessen the need for an actual debating process to take place (c), since we may be able to extract parts of a debate, or entire debates, from existing text; this text need not have been intended to provide arguments on a certain topic, but may contain them anyway. In turn, debating appears to be a necessary process to allow us to develop Argumentation Mining applications, in the first place (d). Without access to existing debates, preferably annotated with various sorts of metadata, we may find it hard to develop formalisms based on which we can build models of arguments with which we can mine text for debates.

Lastly, profiling participants in a debate into different types of users will invariably influence both the way debates are conducted (e) and how they are analyzed (f). Different users in a debate may receive different levels of prominence in a debating interface. The comment of a lecturer in a debate on solutions for a course assignment may be treated as a top level comment in a debate, regardless of the level of participation of other user types, e.g. students or teaching assistants. In much the same way, the comment of a lecturer may be deemed more important than a student's comment when analyzing a debate.

Throughout the remainder of this paper we discuss a number of areas in which we have been developing Argumentation applications. For each application area we discuss whether and how we have addressed the challenges discussed here and, if not, whether and how we may do so in the future.

Application/Use Case	(1)	(2)	(3)	(4)	(a)	(b)	(c)	(d)	(e)	(f)	(5)
Quaestio-it	✓	✓	✗	✗	✓	✓	✗	✗	✗	✗	✗
Desmold	✓	✓	✗	✗	✓	✓	✗	✗	✗	✗	
Argumentation Mining			✓				✓	✓			✓
E-Learning	✗	✗	✗	✗	✗	✗	✗	✗	✗	✗	✗

Table 1. Overview of challenges relating to each application we discuss throughout the paper, where those marked with a checkmark (✓) have been implemented, those marked with a cross (✗) are desiderata and those left blank are not, or only marginally, relevant to the particular application.

3. Quaestio-it: Building a Q&A system

The internet offers a tremendous source of debate and other argumentative structures. One way of both capitalizing on this source and contributing to it is through online debating platforms. To this end we have developed Quaestio-it (www.quaestio-it.com), a web-based Q&A debating platform that allows users to open topics, ask questions, post answers, debate and vote. Below we describe the platform and summarize how we address the challenges identified in section 2.

3.1 Where are we?

In its current state of development Quaestio-it provides an interactive way to engage in conversations on any topic that may be of interest. Through an evaluation algorithm based on the Extended Social Abstract Argumentation (ESAA) Framework (Evripidou & Toni, 2014), the *strength* of answers and comments, as determined by the community, is highlighted. The strength is also visible through visualization in which stronger answers and comments are visibly larger. Within the platform, each answer is open for discussion and users can post their comments, as supporting or attacking arguments, expressing their agreement or disagreement to the answer. Subsequent levels of comments are regarded as attacks or supports to the parent comment/argument. Additionally users can vote arguments of others up or down, increasing or decreasing the strength of the argument voted on. This creates a debate that can be modeled as an ESAA Framework. In order to obtain the relations between arguments within a debate, each user, when posting an argument, has to explicitly state the nature of the comment (i.e. attacking or supporting argument). Answers and comments are then evaluated and the best answer for each question is highlighted. Below we summarize whether and how we have addressed the challenges summarized in figure 1, with the exception of (3), (c) and (d), all of which relate to the mining of debates. We devote section 5 to discuss how this may be integrated with applications such as Quaestio-it.

(1) Enabling debates. In Quaestio-it, debates are represented as trees where (I) the root node corresponds to the initial question, (II) its immediate children correspond to the answers and (III) all other subsequent level nodes are comments (i.e. supporting or attacking arguments on the answers or other comments). One such debate is shown in figure 2. The edges, directed upwards in the tree, connect the nodes to indicate the relations between question, answers and arguments. Dashed edges indicate direct answers to the question, while solid, red (-) or green (+), edges show attacking or supporting arguments on the answers or on other arguments (as posted by the users). Each debate can also be viewed in a more conventional form, which is shown in figure 3. This textual view of the debate includes the ratio of positive and negative votes, strength evaluations and all actions available to the user. These actions are (I) posting a reply and (II) voting positively, negatively or indicating that a comment is irrelevant, malicious or spam. After a predefined number of *spam* votes on an argument or an answer it is removed from the debate, alongside its sub-tree of comments.

(2) Analyzing debates. Figure 4 shows several screen-shots of the development of a debate regarding the recent developments in the political and economical situation in Greece. The debate is about whether or not Greece should stay within the Eurozone.

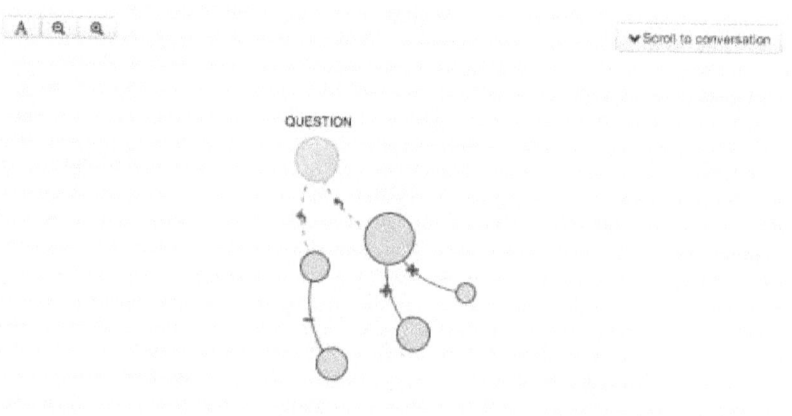

Figure 2. A screen shot of Quaestio-it showing the graphical representation of a debate

Figure 4 (a) shows the initial question posted by the user while Figures 4 (b) and 4 (c) show the two first answers. At this point, both answers have identical strength since none of them have accumulated any votes or arguments. In Figures 4 (d) and (e), two supporting arguments are posted to one of the answers. This increases the answer's strength, making it the best answer for the question. In Figure 4 (f) an attacking argument is posted to the second answer, lowering therefore its strength. Finally, after both answers and arguments have accumulated a number of positive and negative votes, the final state of the debate is shown in Figures 2 and 3 where the best answer is highlighted with a strength of 0.684. This is due to the number of positive votes it has accumulated and the two supportive arguments posted by other users. Nodes vary in size depending on the strength evaluation. This provides a quick insight about the dominant (strongest) answers and comments for a particular question. Hovering over each node displays additional information for each comment or answer including its text and calculated strength.

Figure 3. A screen shot of the application showing the text representation of the final state of the debate

(5) User engagement. Private discussions within the platform can be initiated through the use of *private rooms* where a user, when creating a topic, can select to make it private and send invites to selected users to participate. Each user can create or be invited to multiple private rooms. This was found to be important in order for the platform to be used within organizations and companies wanting to discuss confidential or sensitive information. Section 4 describes such a scenario, where the platform is being used by companies within the injection molding industry to discuss various design decisions and issues.

(a) and (b) There is a reciprocal relation between the way we conduct debates and how we analyze them. For Quaestio-it it was vital to consider what type of graph structure we allow to be generated through a debate. To accommodate the ESAA framework we needed to restrict the debate to tree structures. Since, however, other graph structures may be desirable we will need to consider adapting the ESAA framework in order to fit this purpose, or choosing another framework, altogether. This illustrates how closely challenges (1) and (2) are often linked and we may need to make concessions for one of the two to develop working solutions.

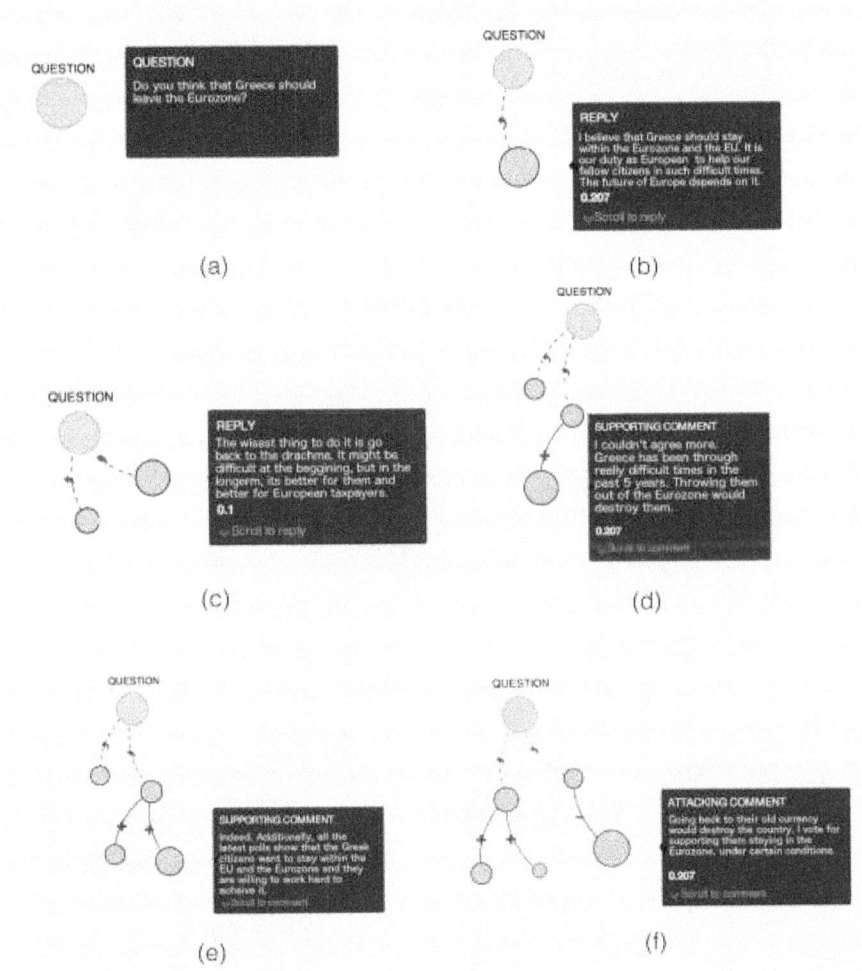

Figure 4. A screen shot of the application showing the development of the debate

3.2 What do we need to do?

Quaestio-it has a number of shortcomings, some of which stem from features that are underdeveloped and other that are absent, altogether. We summarize some below, while we discuss the potential for integration of Argumentation Mining capabilities separately in section 5.

(1) Enabling debates. Through continuous testing of Quaestio-it we have identified some issues in its interface, some of which we have rectified through various updates. Others, however, have yet to be addressed. We have found our choice of

visualization for browsing, shown in figure 5, to be suboptimal. While it works fine as long as the number of topics present is limited, say below 20, finding the right topic becomes more and more cumbersome as the number of topics increases. We hence need to determine a new way of browsing that is both intuitive and remains workable as the topic count rises.

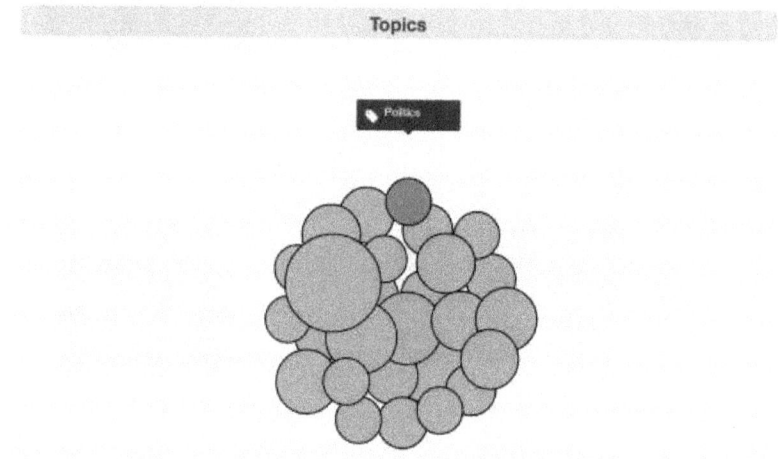

Figure 5. A screen shot of the prototype application showing the topics user interface

(4) User profiling & its integration in (e) the debating process and (f) its analysis. We have discussed the importance of User profiling in section 2, where we point out that, in order to fully capture the meaning and weight of arguments in a debate we may wish to consider who the source of an argument is. In debate forums such as Quaestio-it user profiles may reflect a user's level of participation in certain topics and/or their expertise on it, by whichever way this may be measured. Firstly we may keep track of how active a user is, assuming that the greater the number of contributions on a certain topic the greater their level of expertise. Secondly, other users' opinion on a user, reflected by up and down votes on his/her comments, may be incorporated into determining a user's profile. Lastly we may define predetermined levels of user expertise to, for example, reflect a user's status as some sort of thought leader or expert on a certain subject. A user's profile may then be used to vary the weight his or her comment has in the evaluation of comments' strength values. If, for example, a user is deemed to be an expert on a certain topic and supports someone else's comment on this topic, this comment's score may be strengthened by a larger factor than it would be should a non-expert support it.

(5) User Engagement. Any application may only be considered to be good when one manages to engage and retain users and fosters continuous contribution of said users. To this end, it is often times not sufficient to develop a quality interface that

provides useful functionalities. Another crucial aspect lies in drawing users to the application and keeping them engaged. To increase user engagement for Quaestio-it we intend to take our cues from other social networking applications. The following functionalities will be included:

– Notifications

– Proficiency scores

– Mobile compatibility

Notifications may come in the shape of e-mails or push-notifications, informing users of answers to questions they have posted, comments attacking/supporting their own posts etc. Proficiency scores may aggregate a number of positive metrics, reflecting the level of engagement in the platform a user achieves. A score, for example based on the amount of posts of a user and their scores, can then be displayed to other users, raising one's *standing* in the community (see stackoverflow.com for an example). A major challenge faced by many applications in recent years has been the shift from using applications on personal computers to smartphones. It has become vital to develop applications for multiple platforms, often times making available a desktop application, as well as Android and OS X apps. During the first stage of development we have developed Quaestio-it as a browser application. For it to be adopted by a wider audience, however, we need to render it as accessible as possible. This means that we need to develop applications for all common platforms, not just a browser on a desktop computer.

4. Desmold and Arg&Dec: Building Engineering Applications

Computational Argumentation has proven beneficial in a number of Engineering applications, e.g. (Baroni, Romano, Toni, Aurisicchio, & Bertanza, 2013, 2015; Cabanillas et al., 2013; Evripidou, Carstens, Toni, & Cabanillas, 2014). Expert knowledge can be modeled through Argumentation Frameworks and then analyzed, extracting useful conclusions that can aid engineers throughout the decision making process.

In this section we review two web-based systems to support design in engineering using Argumentation Frameworks, namely the Desmold system (Cabanillas et al., 2013; Evripidou et al., 2014) and the Arg&Dec system (Aurisicchio, Baroni, Pellegrini, & Toni, 2015; Baroni et al., 2013, 2015) (www.arganddec.com). Both systems exploit the power of the web to support collaborative design. They differ, however, in the application area (Desmold is focused on injection molding design whereas Arg&Dec applies to engineering design in general), the underlying Argumentation Framework they use (the ESAA Framework of (Evripidou & Toni, 2014) for Desmold and the QuAD algorithm (Baroni et al., 2015) for Arg&Dec), as well as several other features concerning user engagement.

Below we give further details and analyze how these systems face the challenges we identified in section 2. As in section 3 we leave the discussion of Argumentation Mining integration to section 5.

4.1 Where are we?

Desmold. The *Desmold* project (Cabanillas et al., 2013) (www.desmold.eu) was a collaboration between industry experts in injection molding and Argumentation experts with the aim to build a knowledge-based system to aid injection molding design and prototyping. Within the system, experts can share their experience and opinions, through the debating platform, regarding each design and collaborate throughout the decision-making process. Additional information is provided by a Case Base Reasoner (CBR), in which past debates and designs are stored and retrieved depending on their similarity to the current design being debated. The platform is mainly composed of the following processes: (I) a decomposition process to convert complex geometries into simplified geometries, (II) a debate process supporting argumentation and ontology interoperability to ensure designers' mutual understanding and (III) automatic recommendations based on debates, past experience and rules (see Cabanillas et al., 2013; Evripidou et al., 2014 for details).

To support the debating process, a separate instantiation of the debate component of Quaestio-it, as described in section 3, is used. Each debate is created by the users of the system to initiate a conversation about certain design choices and/or problems. All partners involved in the debate can contribute by providing their opinion in the form of arguments. Supportive material such as documents, images and an ontological representation of the concepts of each design provide additional information to the users to aid them in understanding the given problem and express their agreement or disagreement. Additionally, arguments are constructed from past cases, as provided by the CBR, and incorporated within the debate. Finally, after the debate has finished, the winning answer of each debate is highlighted by taking into account all the available information, in the same manner as in the standard version of Quaestio-it.

(1) Enabling debates. Desmold enables debates of the same form as Quaestio-it (see section 3). Additionally the users have the option to label opinions with concepts from a custom-made ontology for injection molding to add structure to the debate. Users may also expand the debate by retrieving parts of debates from a repository of past cases to be incorporated to extend ongoing debates.

(2) Analyzing debates. Desmold uses the ESAA Framework (Evripidou & Toni, 2014) to analyze debates, taking into account the social support of opinions, as derived by (positive and negative) votes, as well as dialectical relationships of attack and support, as in Quaestio-it.

(a) and (b) Debates vs Analyzing debates. Debates and their analysis are fully integrated in Desmold, in the same way as they are in Quaestio-it.

(5) User engagement. Desmold exploits the functionality of private rooms in Quaestio-it (see section 3) to guarantee privacy and a secure environment for the exchange of information. This allows users to post and discuss sensitive information and grant only a selected group of other users access to the debate. By limiting access to parts of the system through private rooms we hence allow a broader variety of exchanges to take place.

Arg&Dec. The Arg&Dec system (Aurisicchio et al., 2015; Baroni et al., 2013, 2015) supports collaborative design based on debates in a restricted form of the *IBIS* format (Kunz & Rittel, 1972), where issues can be given several answers, which in turn can be debated by providing con-arguments or pro-arguments, in any nested manner. The system is inspired by the *designVUE* system of (*designVUE*, 2013) but it incorporates an algorithm for the automatic analysis of debates that designVUE lacks. Several use cases of the Arg&Dec system (Aurisicchio et al., 2015) and its precursor (Baroni et al., 2013, 2015), in different domains, ranging from wastewater management to construction, have been identified (see Aurisicchio et al., 2015; Baroni et al., 2013, 2015).

(1) Enabling debates. The Arg&Dec system enables debates in IBIS format in the same form as designVUE.

(2) Analyzing debates. The Arg&Dec system uses the QuAD algorithm (Baroni et al, 2015), which requires opinions (issues, answers and arguments) to be initialized with a base score. The algorithm is proven to exhibit a number of properties, including clear relationships with qualitative argumentation (Dung, 1995).

(a) and (b) Debates vs Analyzing debates. Debates and their analysis are loosely integrated in the Arg&Dec system, in that the algorithm needs to be explicitly invoked to determine the strength of opinions in debates.

(5) User engagement. Being targeted at engineers, Arg&Dec also incorporates a matrix-based approach to supporting decision-making, that engineers tend to be familiar with and use in making decisions. Arg&Dec allows to map matrices into debates, and vice versa. It is thus providing a loosely couple integration of two rather different techniques, in an attempt to engage the intended class of users.

4.2 What do we need to do?

Similarly to Quaestio-it, our developments for Desmold and Arg&Dec are work in progress and, as such, certain challenges have yet to be addressed. We summarize the main tasks below.

(3) Argumentation Mining to (c) enable debates Argumentation Mining may offer us ways of incorporating knowledge into debates akin to what we currently achieve through the CBR in Desmold. In addition to integrating past cases from an integrated case base it may be useful to utilize external knowledge bases, such as

Wikipedia or domain specific websites, by mining them for arguments relevant to an ongoing debate. If we are to mine external sources, however, we will need to consider both potential security issues and licensing of information provided by external sources.

(4) User profiling and (e)(f) its integration in the debating process Through the Desmold project we have learned that a myriad of stakeholders may be involved in a debate at any point in time. Engineers, designers, accountants, etc. all may partake in the same debate and contribute from their unique perspective. Profiling users may hence play a decisive role in driving debates to their optimal outcome. A first step in integrating user profiles will be to assign each group of stakeholders certain topics at which they are experts. Whenever such a topic is discussed, a stakeholder's opinion will impact the debate more significantly when he or she is considered an expert on the topic. Other ways of considering users' profiles may include giving them particular rights. If an engineer, for example, argues that a design is technically infeasible, he or she may need to be able to veto arguments, entirely, so as not to consider designs that simply cannot be realized.

(5) User engagement The adoption of a system may be a somewhat smaller worry in a corporate setting than on the web, but users still need to be incentivized to use the system for it to be successful. The main tenant here may be the identification of ways that convince employees that the system eases their work and makes it more efficient. Beyond the generally applicable fact that the interface and functionalities need to be designed in an intuitive and useful manner, there are certain functionalities that we may tailor to the *corporate* setting in which Argumentation is used, here. One way of doing so may be the development of tutorials that give concrete examples of what the system provides and which aspect of one's work it improves.

5. Argumentation Mining

We have introduced Quaestio-it, as well as its adaptation in an engineering setting. In this section we describe a project for which we are using Quaestio-it as a crowdsourcing tool to support the development of Argumentation Mining tools. These, in turn, may offer advanced functionalities in applications such as Quaestio-it, Desmold and Arg&Dec, and many more. For a more in-depth treatment of Argumentation Mining, refer, for example, to Lippi & Torroni, 2016; Mochales & Moens, 2011.

5.1 Where are we?

(3) Mining debates. In Argumentation Mining we are generally confronted with two challenges, (1) the identification of arguments and (2) determining relations between

arguments. These are often treated as separate issues, but we believe that they are intricately linked and hence treat them as one problem. We do this by classifying pairs of sentences occurring in debates according to the nature of the relation that holds between them. The prototype we have developed (Carstens, Toni, & Evripidou, 2014) takes as input two sentences and determines whether one attacks or supports the other, or if it does neither. To classify a sentence pair we represent it as a single feature vector. This feature vector is comprised of two types of features, *Relational features* and *Sentential features*. Relational features represent how the two sentences that make up the pair relate to each other. Features we have been experimenting with on our preliminary corpus include WordNet based similarity (Miller, 1995), Edit Distance measures (Navarro, 2001), and Textual Entailment measures (Dagan, Glickman, & Magnini, 2006). The second category includes features that characterize the individual sentences. Here we are considering various word lists, e.g. keeping count of discourse markers, sentiment scores, e.g. using *SentiWordNet* (Esuli & Sebastiani, 2006) or the Stanford Sentiment library (Socher et al., 2013), and other features. Each of the resulting feature vectors is labeled as belonging to one of three classes, $L \in \{A, S, N\}$, where A = Attack, S = Support and N = Neither. To build a corpus of sentence pairs we have used both Quaestio-it and other sources; here we focus on how we can construct a corpus using Quaestio-it. As an alternative to constructing feature vectors as described above, which can be labour-some, we also consider using simple Bag-of-Words (BOW) representations.

(d) Corpus construction. Quaestio-it offers us a natural resource of sentence pairs that come readily annotated and we can hence use the data collected through the web application to develop models to classify sentence pairs in this manner. We use Quaestio-it to extract pairs of statements of which we know that one either supports or attacks the other. We are thus provided with a set of related pairs that is constantly growing through the use of the Quaestio-it web application. We then use these pairs to test our prototype. What is missing is a set of sentence pairs which are labeled *Neither*. These, however, are easily created. We can simply take posts from Quaestio-it and match them randomly. We then manually confirm that the resulting sentence pairs are actually not related and label them as such. Testing our current prototype on Quaestio-it data we achieve a classification with an F1-score of 0.684.

5.2 What do we need to do?

(3) Continuous development. Argumentation Mining is a fairly young discipline and, as such, we may be some ways away from integrating the described functionalities with applications such as Quaestio-it or Desmold and Arg&Dec. Two of the main obstacles we have encountered concern (I) the construction of corpora and (II) the representation of our sentence pairs via a set of features. Addressing obstacle (I) is, by and large, an issue of investing extensive manual labour. As with any supervised learning problem we need to gather a reasonably large collection of labeled

examples. The definition of *reasonably large* very much depends on the problem we face, but in our case we will need to add, at the least, thousands more labeled examples to the ones we already have. Without doing so it may prove rather difficult to both improve performance and make any kind of judgement on how well our classifications generalize. Obstacle (II) is both an intellectual challenge and, again, one of manual labour. On the one hand we need to identify more features that may reflect characteristics of what characterizes an argument. On the other hand we need to conduct extensive experiments to identify the best combination of whatever features we have at our disposal. We will continue our research to identify features of the two types discussed in section 5.1, those that characterize the relation betweens sentences and those that characterize the sentences, themselves.

(5) User engagement through Argumentation Mining. One of the sources from which arguments are utilized in Desmold is a collection of past cases (Evripidou et al., 2014). These cases are debates that have taken place previously and are similar, by some measure, to the debate currently conducted through the system. The retrieval of past cases is done by measuring similarity with the case discussed. The retrieved debates, or parts of them, can then be inserted manually into the current debate. This can be a rather cumbersome process, as it potentially involves manually parsing both the entire current debate and the debates that have been retrieved alongside the similar cases. Should we be able to perform Argumentation Mining reliably this may allow us to automate the insertion of past debates into the current one, in Desmold as well as other applications.

(c) Enabling debates through Argumentation Mining. Integrating Argumentation Mining with Quaestio-it , Desmold and Arg&Dec may allow for a more fluent exchange of arguments. At present a user needs to consider carefully where and with which type of relation to insert new comments in a debate. If we were able to support or, in a best case scenario, automate this process users could simply post their arguments relating to a topic. We could then establish connections to existing arguments automatically or, in a lesser progression of this development, offer suggestions where an argument may fit into a debate.

6. Towards Argumentation for E-Learning

We have discussed a concrete application of Quaestio-it in an engineering setting in section 4, as well as its usefulness in developing Argumentation Mining solutions in section 5. In this use case we aim to bring together both the concept of Social Argumentation and the automatic identification of arguments and relations to conceptualize an interactive E-Learning platform. E-learning platforms have grown increasingly popular over the past years, with much research devoted to them; see, for example, (Clark & Mayer, 2011) for a recent overview. Knowledge and education resources are accessible to a wide audience. Despite this, conventional E-learning platforms generally fail to reproduce a class environment where students and lectures communicate, exchange ideas and feedback. They are instead focused on lecturers and the likes providing course material that are then *consumed* by the students. Accordingly, much of the focus in such platforms if placed on content, rather than the interaction between all parties that plays a vital role in a classroom environment. We propose an E-Learning platform in which we engage all relevant user groups, e.g. students, lecturers and teaching assistants, by creating an environment that encourages participation rather than sheer consumption of materials. By placing interaction at the centre of the platform we hope to emulate the benefits of classroom teaching in an online environment.

(1) Enabling debates. To allow a free flow of exchange between stakeholders in the platform all interaction can be conducted through an adaptation of Quaestio-it, integrated in a similar manner as in Desmold. This means that posts on exercises, problems, tests, etc. are generally defeasible, encouraging careful reflection on what is posted and whether it is correct/appropriate.

(2) Debate analysis & (4)(e)(f) User profiling. An important aspect of analyzing debates that we have yet to consider is the integration of user profiles. With its clear division of user groups an E-Learning application offers the ideal test-bed for adjusting analysis according to who posts what in a debate. Take, for example a debate on solutions of a coursework. Most likely the majority of posts will be contributed by students who are trying to figure out how to solve the problem. Assuming that the coursework is not trivial (in which case the debate may be rather limited, anyway), there will be disagreement on the approach to the problem, solutions, etc. If, however, the lecturer decides to provide hints or a partial solution, his or her comments should arguably influence the outcome of the debate much more heavily than any student's contribution. Algorithmically, this may, for example, be achieved by assigning a *factor* to each user profile, by which the weight of a post gets multiplied.

(3) Argumentation Mining, (c) its effects on enabling debates and (d) using a new platform to gather data. We hope to integrate input from external sources, such as *Wikipedia*, into discussions by mining them for content that provides fruitful additions to discussions taking place on the platform. This content may come in the shape of arguments when a topic is contentious, and may hence be found through Argumentation Mining as we have described it. Many times, however, a discussion can also be moved along by providing relevant facts that answer open questions or solve problems in a discussion. Here we will need to investigate the applicability of other Information Extraction techniques.

Once we are able to mine external sources for arguments and other input to debates we can utilize our CBR approach from the Desmold system to integrate those contents with debates taking place on the platform. With this approach, debates will generally be comprised of a hybrid of content posted by the users of the platform and content extracted from external sources. All content will be labeled according to attack and support relations and hence offers a new data source. As the platform grows we may thus be able to build ever better models to classify relations, based on a growing corpus of labeled examples.

(5) User engagement. As discussed, building the profile of a user allows us to decide how to judge the impact of their participation in discussions and to assign different rights to different user groups. In addition to that, we may drive user engagement by developing a user's profile according to their level of participation. Many internet fora, such as stackoverflow.com, profile their users based on the amount and quality of their participation. In an E-Learning platform, where discussion is the main driver of participation, this can be integrated seamlessly. A user's profile in the platform will need to reflect the following factors:

– The overall level of participation in discussions
– The quality of participation

The level or participation is a fairly simple measure to attain. Here we mainly need to decide whether we want to opt for an absolute measure, i.e. to just count the amount of posts, or if we should rather measure participation as a proportion of the total level of activity on the platform. Measuring the quality of participation is less straightforward. Through the evaluation algorithm that evaluates the strength of all posts within a discussion, however, we have a natural quality measure at our disposal. The question, then, is whether a simple average of the strength of a user's posts suffices as a quality measure. This will need to be tested once the platform is up and running and we may need to re-evaluate this approach and take other factors into consideration, e.g. test scores of students that modulate a user's profile.

Based on a user's participation scores we can opt for a number of ways to encourage further engagement. Users may obtain badges and increasingly high profile levels. We may also assign certain privileges to users that have a high profile rating. This may entail access to resources not available to everyone, rights to moderate debates, etc.

In an E-Learning platform much engagement will be driven by the lecturers and the degree to which they adapt the platform. If lecturers, teaching assistants etc. drive most interaction outside the classroom through the platform, students who want to gain the best possible results from a course will be incentivized to engage on the platform. We hence need to ensure the ease of use for tasks that lecturers need to accomplish on the platform, such as posting and evaluating tests and questions, providing feedback to students, etc.

(a)(b) Interaction between debate enablement and analysis As with the other systems we have presented in this paper we will need to evaluate how best to integrate the conduction of the debate and its evaluation. We anticipate that the different roles that users can have in an E-Learning platform may need to play a more pronounced role in both the debating process and its evaluation than is envisioned for the other systems. This is the case because the influence of some user groups, e.g. lecturers, on certain discussions, may need to both exert larger influence on the scoring of arguments in a debate and said arguments may need to be highlighted over other arguments. This may be necessary because, for example,

arguments made by a lecturer can be of more interest to students when it comes to answering questions on lecture notes, and students may want to find and access such arguments as easily as possible.

7. Conclusion

Argumentation has the potential to help building useful applications in a multitude of areas. To this end we have presented a number of application developments in order to illustrate the challenges we face when developing concrete Argumentation applications. Based on our experience we have identified five broad challenges, (1) the enablement of debates through good interfaces and applications, (2) the rigorous analysis of debates, (3) Argumentation Mining, (4) user profiling and (5) questions of how we may engage users (see figure 1). Though we have not addressed each of these challenges thoroughly in the applications we have presented, we acknowledge that all five seem to have varying degrees of potential to impact the quality of applications. Based on this we have, for each application, presented a summary of developments that we need to tackle to further the progress of application development in Argumentation.

We believe that it is vital to use Argumentation formalisms out in the real world to give credit to their viability beyond artificial examples and to illustrate the benefits they can provide. This is a process that forces us to constantly re-examine both the applications we develop, as we have done here, but also to do the same with the formalisms we mean to build applications on. Take, for example, the Extended Social Abstract Argumentation Framework (Evripidou & Toni, 2014). We use this framework to calculate the strength of arguments within a tree structure in Quaestio-it. As it stands, however, this framework does not allow us to take user profiles into account when calculating this strength. It is also not trivial to decide whether an attacking comment should be more detrimental to an argument strength than a negative vote on the same argument. We hence need to engage in a process of development in which we not only develop new functionalities, but consider underlying formalisms and the need to adapt them, as well.

Acknowledgments

This research was supported by the EU project *DesMOLD (FP7/2007-2013-314581)*.

References

Aurisicchio, M., Baroni, P., Pellegrini, D., & Toni, F. (2015). Comparing and integrating argumentation-based with matrix-based decision support in Arg&Dec. In TAFA 2015 (pp. 1-20), Springer.

Baroni, P., Romano, M., Toni, F., Aurisicchio, M., & Bertanza, G. (2013). An argumentation-based approach for automatic evaluation of design debates. In CLIMA XIV (pp. 340-356), Springer.

Baroni, P., Romano, M., Toni, F., Aurisicchio, M., & Bertanza, G. (2015). Automatic evaluation of design alternatives with quantitative argumentation. Argument & Computation, 6(1), 24–49. Bench-Capon, T. (2002). Value based argumentation frameworks. In NMR (pp 443-454).

Buckingham Shum, S. (2008). Cohere: Towards web 2.0 argumentation. In COMMA (pp. 97–108), IOS Press.

Cabanillas, D., Bonada, F., Ventura, R., Toni, F., Evripidou, V., Carstens, L., & Rebolledo, L. (2013). A combination of knowledge and argumentation based system for supporting injection mould design. In CCIA (pp. 293-296), IOS Press.

Carstens, L., Toni, F., & Evripidou, V. (2014). Argument mining and social debates. In COMMA (pp. 451-452), IOS Press.

Chesñevar, C., Modgil, S., Rahwan, I., Reed, C., Simari, G., South, M., Vreeswijk, G.& Willmott, S. (2006). Towards an argument interchange format. The Knowledge Engineering Review, 21(04), 293-316.

Clark, R. C., & Mayer, R. E. (2011). E-learning and the science of instruction: Proven guidelines for consumers and designers of multimedia learning. John Wiley & Sons.

Dagan, I., Glickman, O., & Magnini, B. (2006). The PASCAL recognising textual entailment challenge. In Machine learning challenges. evaluating predictive uncertainty, visual object classification, and recognising tectual entailment (pp. 177–190). Springer.

designVUE. (2013, 02). Retrieved from http://www3.imperial.ac.uk/designengineering/tools/designvue.

Dung, P. M. (1995). On the acceptability of arguments and its fundamental role in nonmonotonic reasoning, logic programming and N-persons games. Artificial Intelligence 77(2): 321-358.

Dung, P. M., Kowalski, R. A., & Toni, F. (2009). Assumption-based argumentation. In Argumentation in Artificial Intelligence (pp. 199–218). Springer.

Esuli, A., & Sebastiani, F. (2006). Sentiwordnet: A publicly available lexical resource for opinion mining. In LREC (Vol. 6, pp. 417–422).

Evripidou, V., Carstens, L., Toni, F., & Cabanillas, D. (2014). Argumentation-based collaborative decisions for design. In 26th IEEE international conference on tools with Artificial Intelligence, ICTAI 2014 (pp. 805–809). IEEE.

Evripidou, V., & Toni, F. (2014). Quaestio-it.com –a social intelligent debating platform. Journal of Decision Systems. 23(3): 333-349.

Ghosh, D., Muresan, S., Wacholder, N., Aakhus, M., & Mitsui, M. (2014). Analyzing argumentative discourse units in online interactions. In Proceedings of the first workshop on argumentation mining (pp. 39–48).

Green, N. L. (2014). Towards creation of a corpus for argumentation mining the biomedical genetics research literature. In Proceedings of the first workshop on argumentation mining (pp. 11-18).

Houngbo, H., & Mercer, R. E. (2014). An automated method to build a corpus of rhetorically-classified sentences in biomedical texts. In Proceedings of the first workshop on argumentation mining (19–23).

Kunz, W., & Rittel, H. W. J. (1972). Information science: On the structure of its problems. Information Storage and Retrieval, 8(2), 95-98.

Lawrence, J., Bex, F., Reed, C., & Snaith, M. (2012). AIFdb: Infrastructure for the argument web. In COMMA (pp. 515–516), IOS Press.

Lippi, M., & Torroni, P. (2016). Argumentation mining: State of the art and emerging trends. ACM Transactions on Internet Technology (TOIT), 16(2), Article No 10.

Miller, G. A. (1995). WordNet: a lexical database for English. Communications of the ACM, 38(11), 39–41.

Mochales, R., & Moens, M.-F. (2008). Study on the structure of argumentation in case law. In JURIX 2008 (pp. 11–20), IOS Press.

Mochales, R., & Moens, M.-F. (2011). Argumentation mining. Artificial Intelligence and Law, 19(1), 1–22.

Navarro, G. (2001). A guided tour to approximate string matching. ACM computing surveys, 33(1), 31–88.

Palau, R. M., & Moens, M. F. (2009). Argumentation mining: the detection, classification and structure of arguments in text. In Proceedings of the 12th international conference on artificial intelligence and law (pp. 98–107).

Park, J., & Cardie, C. (2014). Identifying appropriate support for propositions in online user comments. In In Proceedings of the first workshop on argumentation mining (pp. 29-38).

Reed, C., & Rowe, G. (2004). Araucaria: Software for argument analysis, diagramming and representation. International Journal on Artificial Intelligence Tools, 13(04), 961–979.

Socher, R., Perelygin, A., Wu, J. Y., Chuang, J., Manning, C. D., Ng, A. Y., & Potts, C. (2013). Recursive deep models for semantic compositionality over a sentiment treebank. In EMNLP 2013 (Vol. 1631, pp. 1642-1654).

Walton, D. (2013). Argumentation schemes for presumptive reasoning. Routledge.

Wyner, A., & Bench-Capon, T. (2007). Argument schemes for legal case-based reasoning. In JURIX (pp. 139–149), IOS Press.

Wyner, A., Mochales-Palau, R., Moens, M.-F., & Milward, D. (2010). Approaches to text mining arguments from legal cases. In Semantic Processing of Legal Texts: Where the Language of Law Meets the Law of Language 2010: 60-79 Springer.

Wyner, A., Schneider, J., Atkinson, K., & Bench-Capon, T. J. (2012). Semiautomated argumentative analysis of online product reviews. In COMMA 2012 (pp. 43–50), IOS Press.

Chapter 6

How e-Minorities Can Argue on Line. The Case of Teatro Valle Occupato.

Francesca D'Errico

Uninettuno University, Psychology Faculty, Rome, Italy, f.derrico@uninettunouniversity.net

Abstract. This contribution aims at the understanding of argumentative moves in a minority discussing on line within the context of media-activism. The case is represented by some actors of the "Teatro Valle Occupato" (Occupied Valle Theater) in Rome. In this case activists use social media as a complementary tool to their actual protest, thus exploiting "e-tactics". Based on a previous content analysis of verbal change in social media, the present work analyses occupants and followers' argumentations in order to identify the quality of discussions and its possible empowerment effect. Results point out that in general this e-minority is strongly focused on a "normative" and "cultural" core of argumentations, with some differences in "personal" evaluations for what concerns activists. Implications in terms of interpersonal empowerment are considered in the discussion.

1. Introduction

We can start from the notion of discussion. This was considered by most authors as the starting point of several social processes which result in significant outputs, ranging from the construction of knowledge (Billig, 1987; Mosconi, 1990; Mininni, 2003), to the emergence of rules (Sherif, 1958), negotiation between different points of view (Castelfranchi, 1994) and processes of social thinking (Vygotskij, 1978). Discussion and peer cooperation allows the social and moral development and is the basis of processes of social and cultural change.

Groups or minorities that discuss online participate in a common cause in a context that could be defined as *problem solving*, i.e. they should try to harmonize the opposing points of view to achieve a common goal. To do this, they inform, listen and discuss first of all to construct knowledge that is functional to the resolution of the problem, or at best to develop functional but also creative and innovative cultural models.

One of the weapons by which individuals are able to do so is the creation of norms and knowledge through exchange of information and discussion. Discussion includes the clarification of similar or opposing points of view, supported by different types of arguments that can offer the possibility to arguers to impose their position, often influencing bystanders too, but most typically to create the conditions for overcoming the problem, through more or less creative solutions.

In particular, therefore, the defining features of the various types of interaction are the type of group that interacts (*horizontal, vertical, representing the majority or minority, mixed or homogeneous, same goals, conflicting goals*) and the goal to be reached (*interdependent, competitive, collaborative, conflicting*). The intersection of these characteristics gives rise to several forms of discussion and necessarily different decisional output (*integration; compromise, irresolvable conflicts, emancipation and empowerment*) (Castelfranchi, 2015; Pruitt and Carnevale, 1993; Zimmermann, 1988).

The emancipation and empowerment outcome is possible when the discussion is characterized by underlying processes of social support (Bruner, 1986), exchange of information, construction of criticisms and proposals, and when the discussion is strongly influenced by the way of communication and by its quality (Habermas, 1984). The quality may depend on several factors: the level of reasonableness, the critical awareness, but also the creative possibilities given for the problem solving. For the first two characteristics we refer to the notion of "rational-critical debate", according to which the discussion is such if it meets the criteria first of *reciprocity and reflexivity,* in that it is necessary to reflect on the other's positions and actually respond to the questions raised by the other party (Habermas, 1975; Graham, 2003). For the creative level instead we refer to the so called "process of conversion" of active minorities, which triggers precise alternatives and original thoughts (Moscovici, 1981).

One methodological option to identify the quality of communication is certainly argumentation (O-Keefe and Jackson, 1995). One aim of this paper is to try to understand if and how the discussions and arguments of a minority engaged in a real experience of sharing through social media can start a process of participation and personal, social and political empowerment. The constructive strategy of reasoning, arguing and even deciding can be applied on social media too. Social media interactions can produce different effects – in terms of empowerment or active participation - if we consider the type of social media but also the type of group (ingroup or outgroup minority, with or without power; Worchel et al., 1994) that can be solely online, offline but also mixed. In particular, the aim of this work is to explore if and how mixed mediactivism (e-tactics) of an *outgroup minority* without decisional power can construct reasonable, critical but also creative and effective argumentations, bearing in mind that their quality can also elicit empowerment processes. The case we will consider is the "Teatro Valle Occupato" (Occupied Valle Theater) in Rome: a theatre that was occupied by 50 performing artists primarily because local administrators wanted to privatize it. These artists started to organize shows, meetings and discussions inside the theater in order to involve the citizens and to raise funds for the creation of a foundation called "Valle Bene

Comune" (Valle Common Good), as a first example of a cultural site managed as a common good available to shareholders and the Roman citizenship. But what are the features of participatory media?

2. Participative Social Media and Their Features

Social media and social networking seems to be a new way to construct meanings and to build new forms of relationships away of or in support to "real life". From this perspective social media can be used as the main or unique tool to be in touch by means of what we can name "social mediated communities": a virtual togetherness of members that produce discussions and proposals though not in physical presence or by regular contacts but sharing a social category (Tajfel and Turner, 1979) and/or place identity (Dixon and Durrheim, 2000).

To explore the dynamics working within social media and/or communities it seems useful to specify what features can influence their communication. The first feature we posit is the ***type of social media***, i.e., whether the group is *friendship based* (i.e. Facebook) or *topic based* (i.e. Twitter): in the former, users potentially tend to construct their self-image in different ways, more "public" and with a higher level of self-disclosure; in topic based ones users are generally intended to reply to the specific topic or person.

Within these two types one should distinguish different ***types of interaction***: *group (distinguished by accessibility: open, closed or secret)* vs. *one-to-one (private messages) interaction*, and *vertical* (where an influential leader is acknowledged, as in the case of a political leader's fanpage) vs. *horizontal type of interaction* (i.e. group of interests). *Vertical interactions* can be a way to express agreement or disagreement to one person's point of view. For example in political cases the group may simply replicate the leader's way of expression, say, one characterized by aggressive verbal communication (D'Errico et al., 2014); just when the leader exaggerates a polarization of comments may be observed: some tend to the extreme negative pole, but others to the opposite direction; a normative reaction of the followers oriented to stigmatize the leader's verbal aggression. *Horizontal interactions* can instead promote "discursive participation" (Delli Carpini and Cook, 2004), a sort of communication that can promote the use of logic, reasoning and argumentative and critical conversation instead of power dynamics and consequent supine replication to an acknowledged leader (Stromer-Galley and Wichowski, 2010). Horizontal interactions - under certain conditions - for this reason can increase civic engagement in community and democracy (Conover and Searing, 2002).

Both vertical and horizontal interactions can be either *formal or informal*; formal when groups correspond to acknowledged institutions like political parties or formal associations; informal when outside the social network the group is not acknowledged as such. The formality level of the interaction can influence the ways of expressing comments.

Interactions can also have different *goals* (***goals of interaction***) that can affect on line conversation. Based on argumentation theory we can recognise 7 main goals: inquiry, discovery, information seeking, deliberations, negotiation, persuasion, eristic (Walton and Krabbe, 1995). To this Halpern (2012) adds the different levels of emotional involvement (high or sensitive issues) that can compromise, for example, deliberation or persuasion interaction goals.

Another distinction worth consideration are ***participants' features***: they can be inferred from the social indexes contained in the personal profile's description (gender, occupation, personal interests, group memberships, and so on) or they can be deducted from the communicative context or previous knowledge. Lea and Spears (1994) in this connection, with their SIDE theory (Social identity Model of De-individuation Effects) remind that in digital communication it is possible to depict oneself in terms of personal vs. social identity: therefore a person can maintain a relatively fixed and stable personality within the various "social roles" (son, partner, friend, employee, etc ...) s/he can "impersonate" every day even when online. From this *personal or social identifiability,* or on the opposite side *anonymity* (Papacharissi, 2004) de/regulated verbal behaviours may derive: the more participants interact while being de-individuated, the more they communicate impolitely, with very few arguments or with mostly negative evaluative sentences; this is true especially in social networks as youtube - an user-generated online video platform through which individuals can upload and share videos and where users are not required to disclose personal data to log in - differently from Facebook (Halper and Gibbs, 2012).

Within the level of identifiability a person can describe him/herself in terms of different social categories (black, white, men or women, manager or workers, low or high status and so on) and then in terms of different levels of social perception or identity: part of majority or minority (Moscovici, 1996; Mucchi Faina, 2013), ingroup vs. outgroup (Tajfel and Turner, 1979; Turner et al.,1987) or dominated vs. dominant (Lorenzi Cioldi, 2009; Poggi and D'Errico, 2010), against vs. pro certain ideas or associations, groups.

Finally we mention the different ***level of interaction and knowledge*** (just online, offline and mixed); in terms of this feature we may distinguishing different forms of participation and media activism that will be better explained below.

2.1. The "Healthy" Participative e-Minorities Within Social Media

Not all mediated communicative exchanges within social media are participative. We can start from the definition given by Nelson and Wright (1992) whereby participation is "a form in which individuals engage in actions as members of a group with the aim of improving on their conditions", but also a "jointly shared and conscious action for a common cause based on a critical and *conscientization process"* (Friere, 1970) as emphasized in Orford (1992). Jochelovich and Campbell (2000) also identify three fundamental dimensions for what concerns the psycho-social construct: (1) Common social identity and social identification (with a social category), (2) Shared social representation of social context (or world view) (3) Shared knowledge of power relations.

Sharing a social category, a representation of social context and power relations thus seem basic elements and can be framed as dimensions that enable you to understand more in depth other related phenomena such as the "media-activism." Media activism is a particular kind of activism that is undertaken through new media. Considering the communication channel it appears crucial to distinguish the different degrees of media activism, since beside the chosen media it is important to consider the "offline" component. To this purpose within the political domain three distinct types have been recognized (Earl and Kimport, 2011): (1) *E-mobilization*: the web facilitates the sharing of information in the service of the offline such as appointments or information on studies to share. (2) *E-tactics*: this includes online / off line communication; an example is that of "Teatro Valle Occupato", the subject of this paper, in which discussion is present both online and offline, since activists and citizens meet together in the theatre to start a discussion on "theatre as common good", while online they continue to comment and discuss but also try to include those who could not be present at the event. (3) *E-movement*: communication exclusively online. In the case of e-movements, activists do not know each other and they never meet except through social media, they are present in the "groups" only with a common interest in a cause but in different places, they interact just to express their opinions on line.

A question arises looking at these relational and contextual features: what do they imply in terms of empowerment of their participants?

Siddiquee (2006) is one of the first authors, within social mediated communities research, that define social media as a "safe place" because of their informative and supportive functions when, by means of narratives, they provide a way to cope with solitude and isolation. Also Lášticová (2012) analysed social media in relation to immigrants' narratives by pointing out how they use social media as a separate network, different from everyday life, in order to their own wellness.

Authors like Zuniga (2012) already worked on the motivational force of people on their social capital, civic engagement and offline/online participation when they use social media.

Media use and informal discussions are correlated to production of social capital, individual participation in civic and political causes from an informational point of view; this means that media users are more informed and then more active, but when they interact by means of discussions, they comment and at the same time they try to decide a strategy to solve problems; in this case, what are the psycho-social consequences in terms of empowerment?

Before understanding these aspects we need to identify the different forms of empowerment.

Empowerment is defined as a form of re-appropriation of self-efficacy and confidence in the individual or collective range of possibilities to choose (Zimmermann, 1988).

In literature several forms of empowerment are recognized in relation to the levels involved: (1) *individual*, (2) *interpersonal* and (3) *behavioural* with their processes and outputs (Zimmermann, 1988): (1) in the first case the process of community participation leads to results such as individual control of situation, mobilization skills, growing competences and self-efficacy, (2) in the second - the interpersonal level - the process of collective decision making produces organizational networks, and finally (3) at the behavioral level through collective action for access to public resources, the corresponding output will be coalitions, effective resources.

In light of these psycho-social dimensions, recent studies (D'Errico et al., 2015) have highlighted some of those empowering functions starting from social media discussions of Italian researchers (Roars Group on Facebook); authors have shown that Italian researchers protest against the recent university reform also constructing shared strategies by means of shared decisional making on common or individual actions, and that this way of interacting was a typical interpersonal empowerment for participants.

Recently D'Errico (2016) analysed the forms of empowerment put in place in relation to a particular type of mediactivism: an "e-movement" built by an "ingroup minority" (an Italian network of researchers, Roars). This study has highlighted such as "e-movement" is a way to talk about norms, rules useful to overcome a critical phase and to promote a strong form of interpersonal empowerment, but not a political one.

3. The Quality of Minority Participation: How They Argue.

One of the basic objectives of this study is therefore to see how one of the possibilities to identify and promote empowerment is just to promote a high quality discussion.

The features of "a good discussion" have been identified through various contributions with reference to the subject of deliberation. However (Habermas, 1975; Dahlberg, 2001; Graham, 2003) they start from the notion of "rational-critical debate" which highlights as "no force except that of the better argument is exercised" (Habermas, 1975: 108). So, good argumentations can play a central role in a debate, but what are the features that make them qualitatively high?

Following Graham (2003) argumentation can drive a rational critical debate, it requires that participants provide reasoned claims, which are critically reflected upon, it requires *coherence, continuity* (stay on the topic) *reciprocity* (participants listen and respond to each other questions and arguments), *reflexivity* (an internal process of reflecting on another participant's position), *perspective taking* (understanding and conceptualization with empathic mood).

These considerations push to better understand the issue of quality in communicative exchanges and arguments used by minorities discussing online.

The link between active minorities and their rules of argument has been pointed out just within an experimental approach, underestimating the role of argumentation from a qualitative point of view (Mucchi Faina et al., 2013).

First of all, the argument is an activity-oriented verbal behavior aimed at convincing a reasonable critic of the acceptability of a standpoint (Van Eemeren, 2003; Walton et al., 2008); generally the arguers tend to justify the acceptability / positivity of their views using arguments in their favour, and to reject opposing viewpoints (confirmation bias, Kuhn, 1991). If this is the main objective of the arguers, from a cognitive point of view, the tendency would be one to confirmation rather than one to refusal (Kuhn, 1991).

In this regard, Mucchi-Faina and Cicoletti (2006) have shown that a minority promotes greater quality of cognitive processing compared to majority, since it promotes conversion, alternative and original thoughts, differently from the majority who instead promotes validation processes or compliance (Moscovici, 1981).

Within the rhetoric and argumentation field, and starting from a pragma-dialectic perspective, Walton and colleagues (2008) have identified 60 types of "argumentation schemes" that can for example be based on both confirmations - such as generally accepted opinions, bias, ignorance – and on possible disconfirmation - as in the case of cause-effects, evidence or consequence schemes.

Arguers, in Walton's (2008) description, can choose from different types of argumentation schemes, in the majority based on "presumptive" ways of reasoning, because arguments are used mostly when there is an absence of evidences or facts, so arguers need to explain available evidences by making persuasive inferences. Therefore, he acknowledges *descriptive argumentation schemes* that give some hints starting by the source of information as in the case of *argumentation from evidences, memoriam, perception, from ignorance,* or the ones that come from emotions, as in the case of *argumentation schemes from threat, fear appeal, danger appeal, distress.* Walton, Reed and Macagno (2008) list also *normative argumentation schemes* where sentences or arguments are based on norms, rules or other positive or negative examples, analogies, authorities or experts. Within these are *argumentation schemes from rules, examples, analogies, ethotic argument.* The normative argumentation can be differentiated from the *presumptive argumentation schemes*, based instead on possible inferences as in the case of *abductive argumentation, practical reasoning or pragmatic inconsistency, from consequences, from sign to consequences, slippery slope argument, from alternatives, from oppositions, from evidences to a hypothesis, cause to effects, correlation to cause*. Other argumentation schemes are based on the classification and specification of verbal features used in the argumentation as in the case of *vagueness or arbitrariness of verbal classification* that allows a proper contextualization of the arguments.

In particular the act of discrediting the other to demonstrate he is wrong may be sometimes a case of *"ad hominem fallacy"*. It is a discrediting move aimed at damaging the other's image (D'Errico and Poggi, 2012) that violates the "freedom rule", according to which participants in a discussion must be free to provide arguments without fearing of being attacked.

Furthermore Van Dijk (1992) points out how the argumentative approach has a double core: "structural" and "functional". The structural approach concerns cognitive strategies aimed at persuading, that can be oriented to the description of what or how someone has heard or seen, they concern logical (truth preserving), psychological (plausibility preserving) or social (interactionally relevant, normative) inferences. (p.246) So they can be true, plausible or normative.

At the same time Van Dijk (1992) recognizes broader social, ideological or cultural functions of argumentations that consist in the fact that the arguer can be a member of a group and give his discourse biased or very biased argumentations. He can also give an admitted or non admitted evaluation, as in the case of argumentation from a personal point of view (argumentation from *bias, from position to know, expert opinion, witness testimony, or ad hominem, generic ad hominem and circumstantial ad hominem, that represent* particular violations to argumentation schemes*)*.

This can be a first way of acknowledging a functional aim; the second, as asserted by Van Dijk coherently with his socio-cognitive approach, is the fact that argumentations can be expression of a more social, cultural or ideological point of view, that is, in terms of group membership ("from the group to its member"), they may stem from a value, popular opinion, popular practise.

Following Walton's and Van Dijk's considerations we can describe argumentation schemes in the table below:

Argumentational core	Level of argumentation	Types of argumentation scheme
Structural	*Descriptive*	Evidence, memoriam, perception, ignorance, threat, fear, danger, distress
	Presumptive	Abductive, pratical reasoning, pragmatic inconsistency, from consequences, alternatives, oppositions, from evidences to hypothesis, cause to effects, correlation to causes
	Normative	Rules, examples, analogies, commitment, ethotic argument
Functional	*Personal evaluation*	From an expert, witness testimony, from a bias, ad hominem, generic ad hominem, circumstantial ad hominem
	Social and cultural evaluation	From the group to its member, from values, popular opinion, popular practise, position to know

Table 1. Argumentational core, level of argumentation and types of argumentation schemes

Taking a rhetorical view into consideration, the present work explores a psycho-social issue: what means to be part of a minority group in case of e-tactics, that is when participants interact both offline and online; what discourses and argumentations are used to face a problematic phase against a public institution (municipio, comune di roma = Roma city hall) in which participants are involved, going beyond confirmation or falsification as type of cognitive processing.

Further, if minority promotes creative cognitive processing and improves its quality (Moscovici, 1981; Mucchi Faina et al. 2013) which argumentations, what thoughts does it favour, among the wide range of them, especially under critical conditions? And when do these promote empowerment processes to face those conditions?

4. The Case of Teatro Valle Occupato

The case presented here concerns a particular e-minority: an *outgroup minority*, without decisional power. They are activists within the frame of *"e-tactics"* because they use social media both offline and online, to promote civic engagement and participation. The case is one of "Teatro Valle Occupato".

Teatro Valle is one of the oldest theaters in Rome, with history dating back to 1727, and it operated as Italian Theatre Institution (ETI= Ente Teatro Italiano) until May 2010. ETI was a non-profit organization with the aim to promote and disseminate theater activities following ministerial directives. After the closing of ETI, the Mayor of Rome Gianni Alemanno, belonging to a leftist party, decided to make it a private institution. On June 14, 2011 Teatro Valle was occupied by a group of 50 performing artists, activists and free citizens who protested to have such an important theater remain a public space and be managed in a transparent manner. Their slogan was "How sad it is to be prudent", quoting the Argentinian playwriter Rafael Spregelburd, and it has become a motto of the movement (http://romethesecondtime.blogspot.it/2011/10/occupy-theater-teatro-valle-occupato-in.html). Since the day of the occupation, actors and activists have started to manage the theatre in organizational aspects but also in entertainment programming, involving a huge number of famous artists, actors and singers that have in their turn involved ordinary people; this fact rose to the national news as a first example of "theater as a common good". At the same time they launched a public discussion on what should be the most appropriate legal form of management of the theatre as a "common good".

This discussion, along with fund raising among ordinary citizens who participated in the demonstrations, led to the "Declaration of Teatro Valle Bene Comune" held on 14 June 2014, on the basis of processes of group thinking and discussion on the "commons" with lawyers and political figures.

So in this occupation we can recognize a first stage of improvised protests and recognition of the means of subsistence for the theater's management, a second one characterized by progressive involvement of Citizenship through performances, meetings and social media - Facebook and Twitter - up to the third one where a first legal acknowledgement occurred of the theater as "Fondazione Teatro Valle Bene Comune".

4.1. Procedure, Corpus and Lexicometric measures

We collected discussions from the blog www.teatrovalleoccupato.it, where occupants created a discussion space called "Agorà" from June 2011 to April 2014. We obtained a corpus which counts 139426 (V) occurrences with 20710 (N) different words and a medium lexical richness index [(V/N)*100], equal to 14,85% (that is under 20%, reference point of the corpus richness; Bolasco, 2013). An automatic quanti-qualitative analysis was performed on the occupants' and protesters' comments by TalTac (Trattamento Automatico Lessicale e Testuale per l'Analisi del Contenuto, i.e. "Lexical and Textual Automatic Processing for Content Analysis": Bolasco, 2013), a software for textual data analysis based on a "lexicometric approach": an application of statistical principles to textual corpora.

The lexical analysis includes some descriptive information, particularly interesting for understanding the conversational dynamics of Teatro Valle occupants and their followers, like time and person and adjective analysis (*Imprinting analysis*).

The corpus is oriented toward present time (64%), the person used is the first plural person "We", while the adjectivation is positively oriented (58%).

4.2. Teatro Valle's Topics and Argumentations

4.2.1. Semantic Analysis of Teatro Valle's Forum.

In a previous study (D'Errico et al., *2015*) we analyzed the peculiar and characteristic lexicon of Teatro Valle's occupants to see their participation dimension (Campbell and Jochelovich, 2000) and to investigate the role played by interactions within the forum. By means of most frequent words and their real use within the real sentences (concordance analysis), three main categories have been extracted named as follows: 1) *a common identification in a social category* 2) *a shared representation of social context* 3) *a shared knowledge of power relations*.

The first category was named "common identification in a social category", where activists try to form a common (critical) identity passing through different stages. With respect to four different years analysed (2011-2014) they move from a "critical consciousness" in the first year of occupation (2011: frequent words like *entertainment workers, critical action of the company, strike, critics, models, reappropriation, dignity*), to a more "planned form of networking" (2012: frequent words, *collaboration, collective, participation, networking, strategy, design, social comparison, democracy, public assembly, meeting,* along with *performance, appeal, common*) and "permanent space and collective/creative actions" (2013; frequent word, *seasons, committee, theater, contemporary, energy, expression, knowledge, creative committee, representation, artistic, theatral season*) to a "definition of legal and acknowledged form: Teatro Valle Foundation" (2014; frequent words: *institutions, laws, teatro valle foundation, Rodotà, overspend, property, territory, city, perspectives, government practices, initiative, principles, rights, recognition, legislation*).

Furthermore, the second and the third category point out how Teatro Valle occupato's forum represents a space where people discuss on socio-economic problems that affect their precarious conditions by looking also at the possible legal alternatives to react and improve the situation. As to the economic context, activists, in interaction with their followers, focus on the lack of investments in the cultural and artistic field, unbalanced legal protections, and try to understand the way to gain some rights, for example by appealing to "third institutions" and to work on "*commons and rights*" with jurists recognized by the institutions (*Rodotà, Mattei, Ostrom*). Finally they use the forum to organize themselves in order to make concrete actions (the most frequent words: *claim, re-appropriation, struggles, recover, expropriation, strike, self-government, take back, space, occupy, rebel*).

Summing up, according to the previous study the forum allows to (1) form a common collective identity, (2) seek socio-economical causes, creative solutions and legal and alternative stakeholders to react to the critical condition, 3) concretely and actively organize to put them in place.

4.2.2. *Argumentation Analysis of Teatro Valle's Forum*

We start from the extracted lexicon in the previous study (D'Errico at al., 2015) for argumentation coding following Walton's (2008) schemes previously overviewed. The descriptive percentages below point out that normative argumentations are dominant within the whole corpus (37,7%) followed by cultural evaluation (22%) and personal evaluation (20,2%). In particular while activists' argumentations are focused on normative (34,1%), personal (24,1%) and cultural evaluation (21,8%), followers' ones are strongly oriented toward normative (43,8%) and much less on the other level (22,3% cultural evaluation; 13,2%personal evaluation).

	Descriptive	*Normative*	*Presumptive*	*Personal evaluation*	*Cultural evaluation*
Follower	10,7%	43,8%	9,9%	13,2%	22,3%
Activists	19,9%	34,1%	0,1%	24,1%	21,8%
Total	16,6%	37,7%	3,6%	20,2%	22,0%

Table 2. Level of argumentation*sender type

In a sense, while followers stare at the structural core when they argue, activists alternate structural to both personal and cultural evaluation, so looking at the ideological and functional core (χ^2 (4)= 31,33; p<0.0025).

In particular, when arguing, occupants focus mainly on the "normative core" with high frequency *argumentations from rules* (19.4%) and with less frequent argumentations *from committment* (9.5%) and *example* (5.2% ; Table 3). Next to this, occupants express their personal or others' point of view by using different voices with either "bias", "expert opinion" and "position to know". They also try to ground their occupation by discussing with argumentations from value (10,9%) and to a lesser extent with popular opinion (5.7%).

Argumentational core	*Scheme*	%
Descriptive	Evidence	6,2%
	Distress	5,2%
	Need for help	8,5%
Normative	Rules	19,4%
	Committment	9,5%
	Example	5,2%
Personal evaluation	Bias	9,0%
	Expert opinion	9,0%
	Position to know	6,2%
Cultural evaluation	Value	10,9%
	Popular opinion	5,7%
	From group to its member	5,2%

Table 3. Activicts' argumentation schemes

If we analyze in detail "argumentation from rules", we can recognize four main groups: 1) legal norms and juridirical issue, 2) rules for public opinion involvement, 3) rules for promoting cultural and value changement 4) rules for discussion and deliberation within the forum. Moreover, after those on "normative core", we will list arguments focused on personal and cultural evaluation.

(1) Legal norms and juridirical issue

a. "the *law* literally applied, as said above, would exclude almost all workers in the entertainment... the only hope is to convince a judge to raise the issue of unconstitutionality!"

(la legge applicata alla lettera come detto escluderebbe quasi tutti i lavoratori dello spettacolo . . . unica speranza è convincere un giudice a sollevare la questione di incostituzionalità!)

b. "Among the various legal actions we have **recourses**, amendment to the budget and law proposals to repeal or change the law from '35. In the other meeting I spoke about the recourse that I made to unmask the unconstitutionality of this law"

(Fra le varie azioni, a livello giuridico, ci sono i ricorsi, l'emendamento alla finanziaria e proposte di legge per abrogare o cambiare questa legge del 35 e i ricorsi avverso la circolare che possono fare le associazioni di categoria. Nell'altra assemblea parlai del ricorso che feci per smascherare l'incostituzionalità della legge del 35.)

(2) Rules for public opinion involvement

a. "We should make proposals that could be a platform; seek recognition of our artistic work, require unions and those most in view. We get out two or three proposals on which everyone can agree on: the unions, we, associations"

(Dovremmo fare delle proposte che possano essere una piattaforma; chiedere il riconoscimento del lavoro artistico, richiedere sindacati e dai soggetti più in vista. Tiriamo fuori due o tre proposte sulle quali possono concordare tutti: sindacati , noi , associazioni.)

b. "Promoting a process of information, discussion, mobilization and articulation of proposals on the issues of environmental and social justice and on the need to move towards a model based on justice in Italy"

(Promuovendo un percorso di informazione, discussione, mobilitazione e articolazione di proposte sui temi della giustizia ambientale e sociale e sulla necessità di operare una transizione verso un modello fondato sulla giustizia in Italia).

(3) Rules for promote cultural and value changement

a. "New models of development for a direct management: see "if not now, when" (feminist association) with women the cultural issue has been raised, we are mature. Why couldn't the new generation participate and decide? How possible for the people, associations, groups, to enforce this community?"

(Nuovi modelli di sviluppo per una gestione diretta se non ora quando con le donne la questione culturale sia stata sollevata, siamo maturi. Perchè non potrebbe partecipare la nuova

generazione,e decidere Come possibile per la gente, le associazioni, i gruppi, far valere questa collettività?)

b. "***copyleft*** does not exclude financing systems. Wouldn't it be nice to imagine culture accessible to all, freely and publicly financed?"
(*il copyleft non esclude a priori sistemi di finanziamento. Non sarebbe bello poter immaginare cultura accessibile a tutti, finanziata liberamente e pubblicamente?*)

c. "The will is to establish a new committee made up of lawyers, with the contribution of those who in recent years have worked in contact with the theme of ***commons***, the protagonists of the struggles of the commons and of those parliamentary members who will prove to be sensitive to the need to insert the institution of common godos into the code of property rights."
(*La volontà è quella di costituire una nuova commissione composta da giuristi, con il contributo di coloro che in questi anni hanno lavorato a contatto col tema dei commons, dei protagonisti delle lotte dei beni comuni e di quei parlamentari che si riveleranno sensibili alla necessità di inserire nel codice del diritto di proprietà l'istituto dei beni comuni.*)

d. "We have to change the cultural system to change Italy, we are ready to try. We joined our companions from Venice, we must start bottom-up"
(*Dobbamo cambiare il sistema culturale per cambiare l'Italia. siamo pronti a provarci. Ci siamo uniti ai compagni di Venezia, ma partire dal basso.*)

<u>*(4)Rules for discussion and deliberation within/outside the forum*</u>

a. "The goal would be to make participants more aware, so that at the meeting we can discuss the proposals already well-rehearsed and updated in the forum by various interventions."
(*L'obiettivo sarebbe quello di rendere più edotti e consapevoli i partecipanti, in modo che in assemblea vengano discusse le **proposte** già ampiamente sviscerate discusse e modificate nei forum dai vari interventi*).

b. "The Forum will open a discussion, the various proposals will be reviewed and discussed: turn-over of art directors, division by categories (education, production, programming), choices made with greater awareness in order to vote the most interesting one."
(*Sul forum si aprirà una discussione, verranno esaminate e discusse le varie **proposte**: rotazione dei direttori artistici, suddivisione per categorie (formazione, produzione, programmazione), modalità di scelte con maggior consapevolezza votare la più interessante*).

Argumentation from Bias:

a. "the hegemonic model, that is, the productive one, for which culture would be valued or not depending on its ability to produce goods."
(*il modello egemonico, ovvero quello produttivo, per il quale la cultura avrebbe valore o meno a seconda della sua capacità di produrre merci*)
b. "The policy has produced a vacuum, created patronage and civil servants of this mechanism."
(*La politica ha prodotto un vuoto, creato clientele e funzionari di questo mecanismo*).
c. "on the part of politicians, who want to impose nominations, while bypassing rules and common sense, this requires us to react."
(*da parte della politica, che vuole imporre nomine, scavalcando regole e buon senso, ci impone di reagire.*)

The followers answer by supporting the positions taken by activists, demonstrating a high level of *reciprocity, reflexivity* and even *empathic perspective taking* (Graham, 2003); they relaunch on issues of *values* or on other legal proposals, as in the following case. Activists raise criticalities and followers respond empathically or by proposing a solution, a norm or reaffirming a basic value, as in the following sequence:

> (Occupant): "It's time to change things, starting from the critical issues, to deal with a theater in every town and build a connective tissue, this takes decades, it would be better to change what this country must be thinking in twenty years.
> (Follower): "We are here on purpose" (*argumentation from committment*)
> (occupant): "Sure, but not easy. I am not part of any lobby. I also propose a system of turnover crossed: when an art director is nominated, after a year he has to change the supervisory body, etc. Even this would give more oxygen to the system. One last thing: I admire what you do. "(*argumentation from rules*),
> (Follower): "To start with the concept of cuts: we complain about a lot of these cuts to "culture"; we have to change the thinking behind this. It requires a cultural revolution that can change awareness. In the modern conception, the only serious things are privatized. We must reverse this concept"(*argumentation from value*). [1]

---

[1] (occupant): "*È arrivato il momento di cambiare le cose partendo dalla criticità, si occupi un teatro in ogni città e costruiamo un tessuto connettivo, questo richiede decenni, sarebbe meglio cambiare pensando cosa deve essere questo paese tra vent'anni.*"
(follower): " Siamo qui apposta".

Quantitative data support this trend, highlighting how the followers of the forum use primarily arguments coming from *values* (22.3%; Table 4) and in the second place "argumentation from *rules*" (18.2%); also in this case *commitment* and *examples* are fairly present, (respectively, 10,7% and 14,9%). Arguments that refer to a personal evaluation are relatively infrequent (13.2%), and no "ad hominem" attack was coded. These percentages are significantly different from the ones of the activists (χ^2 (6)= 16,46; p<0.005).

Argumentational core	*Scheme*	%
Descriptive	Evidence	10,7%
Normative	Rules	18,2%
	Committment	10,7%
	Example	14,9%
Presumptive	From evidence to hp	9,9%
Personal evaluation	Bias	13,2%
Cultural evaluation	Value	22,3%

Table 4.Followers*argumentation type

Below will be present some examples in which many followers through forum discussions reconstruct and revitalize the sphere of values, with particular attention to the dignity of workers, both through initiatives that support them, and through mutual encouragement. In this case the "*arguments from value*" help followers in terms of the *psychological and interpersonal empowerment*; it is clear from these concordances how arguing became a way to consolidating and sharing self-awareness of being an artist, worthy of "rights" and "dignity".

5. Discussion

(occupant): "*Certo , ma non facile . Io non faccio parte di nessuna Propongo inoltre un sistema di turnover incrociati: nel momento in cui un direttore artistico nominato, dopo un anno deve cambiare l'organo di controllo, ecc . Anche questo darebbe maggior ossigeno al sistema. Un'ultima cosa : ho molta ammirazione per quello che fate*".

(occupant): *Per partire dal concetto dei tagli: ci stiamo lamentando moltissimo di questi tagli alla cultura; dobbiamo cambiare il pensiero alla base. é necessaria una rivoluzione culturale che consenta di cambiare la coscienza comune. Nella concezione odierna, le sole cose serie sono quelle privatizzate . Bisogna invertire questo concetto*"

The first consideration raised from this work is that "talking about a revolution" – cultural or social – on social media does not mean simply to be against someone, but can be a way of discussing and arguing mainly on group norms and laws.

The analysis of online discussions and argumentations of the "Teatro Valle Occupato" as a form of e-tactics of media-activism made it possible to see how an e-minority can promote a "qualitative" route for a solution of a real problem.

Specifically, the present work has focused on the argumentative dimension of occupants and their followers' on line discussions, in order to extract and analyse the quality of their communicative exchanges. We start from a general framework that states that the quality of argumentation is based on a (1) "socio-cognitive" tendency to create new solutions and creative norms in order to solve a given problem, as in the case of active minorities (Moscovici, 1981), but also (2) on a general level of reciprocity and reflexivity between on line interactants (Graham, 2003) during the discussion. In order to analyse the quality of discussions we used an argumentational approach (Walton, 2008) useful to understand the rhetorical core, of both followers and activists. From the results it emerges that "talking about the revolution" in a e-tactics context could correspond to find new solutions in terms of civil and workers' rights, as activists tend to share and discuss on "rules" (19,4% on the whole argumentation coding) concerning legal norms, rules and proposal for public opinion involvement, rules for promoting cultural and value changement.

On their side, followers support activists by giving their "commitment" but mainly giving other virtuous "examples" close to their experience.

In addition, occupants express their anger for injustice often arguing with personal evaluations, and thus they use many "argumentations from bias" and "from expert opinion", but no "ad hominem argumentation" occurs. In the same context, followers argue by relaunching higher *values* such as one of dignity, using in the majority of the case argumentations from "values" (22,3%).

Thus, *Teatro Valle Occupato* forum seems a good way to construct the change of norms and values and it seems be a good form of *interpersonal empowerment*, because to the problems and negative evaluations proposed by activists a reciprocal support corresponds on potential proposals and renewal of basic values (of "being artist"). In this case the *political empowerment* is acted by means of theater occupations and thus the social media contribute just in the common identification for a real participative actions (D'Errico, 2016).

I want to conclude with a recent thought of Papachirissi (2011: 78) that is in line with the results of the present study, on investigations of the social media participation: "it is possible that our quest for civic behaviors has not produced the desired results because we have not been looking at places that civic behaviors now inhabit: spaces that are friendlier to the development of contemporary civic behaviors".

References

Billig, M. (1987). *Arguing and thinking*. CUP
Bolasco, S. (2013). L'analisi automatica dei testi. Fare ricerca con il text mining. Roma: Carocci.
Bruner, J. (1986). *Actual minds, possible worlds*. Cambridge, MA. Harvard Univ.
Campbell, C. and Jovchelovitch, S. (2000). Health, community and development: towards a social psychology of participation. Journal of Community and Applied Social Psychology, 10 (4), pp.255-270.
Castelfranchi C. (1994). "Ma non dica idiozie!" Per un modello delle interazioni verbali al di là della conversazione. In: F. Orletti (ed), *Fra conversazione e discorso: l'analisi dell'interazione verbale*, La Nuova Italia Scientifica, Roma, 143-170
Castelfranchi, C. (2015). The cognition of conflict: ontology, dynamics, and ideology. In: F. D'Errico et al. (eds) *Conflict and Multimodal Communication* (pp. 3-32). Springer International Publishing.
Conover, P.J., & Searing, D.D. (2002). Expanding the envelope: Citizenship, contextual methodologies, and comparative political psychology. In J. H. Kuklinski (Ed.), *Thinking about political psychology* (pp. 89-114). New York, NY: Cambridge University Press.
Delli Carpini, M.X., Cook F.L., & Jacobs, L.R. (2004). Public deliberation, discursive participation, and citizen engagement: A review of the empirical literature. *Annual Review of Political Science*, 7, pp. 315-344.
D'Errico, F. (2016). With different words: The arguments that can empower an e-minority. *Computers in Human Behavior*, 61:205-212
D'Errico F., Poggi I., Corriero R. (*2015*) How sad prudence is". Teatro Valle Occupato as a case of minority empowerment through media-activism. In Proceedings of SMART 2014 Medimond S.r.l. - Monduzzi Editore International Proceedings Division
D'Errico, F., Poggi, I., Corriero R. (2015). Minority group discussions as resilience strategy in social media. The case of "roars" in the Italian academic context. In *Proceeding Ceur Essem 2015. Emotion and Sentiment in Social and Expressive Media*. Vol-1351, pp. 116-126.
D'Errico, F and Poggi, I., Corriero R. (2014). Aggressive language and insults in digital political participation. In Proceedings of Conference on Computer Science and Information Systems: *Web Based Communities and Social Media 2014*, pp.105-114.
D'Errico F, Poggi I., (2012) Blame the opponent! Effects of multimodal discrediting moves in public debates. *Cognitive Computation*, vol 4(4), pp.460-476
Dixon, J. & Durrheim, K. (2000). Displacing place identity: a discursive approach to locating self and other. *British Journal of social psychology*, 39 (1), pp.27-44
Earl, J. & Kimport, K. (2011). Digitally enabled social change: activism in the internet age. Cambridge, USA: MIT Press.
Freire, P. (1970). *Pedagogy of the Oppressed*. London: Penguin.

Friedmann, J. (1992) Empowerment: The Politics of Alternative Development. Oxford: Blackwell.

Garmezy, N. (1991). Resilience in children's adaptation to negative life events and stressed environments. *Pediatrics.* (20), pp. 459–466.

Graham, T., & Witschge, T. (2003). In search of online deliberation: Towards a new method for examining the quality of online discussions.Communications/Sankt, 28(2), 173-204.

Halpern, D., & Gibbs, J. L. (2013). Social media as a catalyst for online deliberation? Exploring the affordances of Facebook and YouTube for political expression. *Computers in Human Behavior*, 29, 1159-1168

Hamill, A., Stein, C. (2011) D Culture and Empowerment in the Deaf Community: An Analysis of Internet Weblogs. *Journal of Community & Applied Social Psychology*, 21, pp. 388–406

Kuhn, D. (1991). *The skills of argument.* Cambridge: CUP

Lasticova, B. (2012). New media, social capital and transnational migration: Slovaks in the UK. *Human Affairs,* 24(4): 406-422

Lebart, L., Salem, A. (1994). *Statistique textuelle*, Paris: Dunod.

Lorenzi Cioldi F. (2009) Dominants et dominés : Les identités des collections et des agrégats, Grenoble: PU.

Lortie-Lussier, M., Lemieux S., Godbout L. (1989). Reports of a Public Manifestation: Their Impact According to Minority Influence Theory. *Journal of social Psychology*, 129 (3), pp. 285-295

Mininni, G. (2003). *Il discorso come forma di vita.* Guida Editori.

Mosconi, G. (1978). *Il pensiero discorsivo.* Il mulino.

Moscovici, S. (1981). *Psicologia delle minoranze attive.* Torino: Boringhieri

Mucchi Faina, A., Pacilli G., Pagliaro S. (2013). *L'influenza sociale.* Il Mulino: Bologna

Mucchi Faina, A. e Cicoletti, G. (2006). Divergence vs. ambivalence: effects of personal relevance on minority influence. *European Journal of Social Psychology*, 36, 91-104.

Nelson, N., Wright S. (1992). *Power and Participatory Development. Theory and Practice.* Intermediate Technology Publications: London.

O'Keefe, D. J., & Jackson, S. (1995). Argument quality and persuasive effects: A review of current approaches. In *Argumentation and values: Proceedings of the ninth Alta conference on argumentation* (pp. 88-92). Annandale, VA: Speech Communication Association.

Orford, J. (1992). Community Psychology: Theory and Practice. Chichester: Wiley.

Paglieri, F. (2015). Arguments, Conflicts, and Decisions. In: F. D'Errico et al. (eds) *Conflict and Multimodal Communication* (pp. 117-136). Springer International Publishing.

Poggi I. & D'Errico F. (2010) Dominance in political debates. In A.A. Salah et al. (Eds.): *Human Behavior Understanding*, LNCS 6219, pp. 163--174. Springer, Heidelberg.

Pruitt, D. G., & Carnevale, P. J. (1993). Negotiation in social conflict. Thomson Brooks/Cole Publishing Co.

Papacharissi, Z. (2011). A networked self. *A Networked Self*, 304.
Papacharissi, Z. (2004). Democracy Online: Civility, Politeness, and the Democratic Potential of Online Political Discussion Groups, *New Media & Society*, 6(2), pp.259-284.
Reicher, S. D., Spears, R., & Postmes, T. (1995). A social identity model of deindividuation phenomena. In W. Strobe & M. Hewstone (Eds.), *European review of social psychology*, 161-198.
Siddiquee, A., Kagan C. (2006). The Internet, Empowerment, and Identity: An Exploration of Participation by Refugee Women in a Community Internet Project (CIP) in the United Kingdom (UK), *Journal of Community & Applied Social Psychology*, 16: 189–206
Sherif, M. (1958). Group influences upon the formation of norms and attitudes. *Readings in social psychology*, 219-232.
Spears, R. & Lea, M. (1994), Panacea or panopticon? The hidden power in computer-mediated communication. *Communication Research*, 21, pp. 427-559.
Stavrositu, C., Sundar, S.S. (2012) Does Blogging Empower Women? Exploring the Role of Agency and Community, *Journal of Computer-Mediated Communication*, 17, 369–386
Stromer-Galley, J., & Wichowski, A. (2010). Political discussion online. In M. Consalvo & C. Ess (Eds.), *The Blackwell handbook of Internet studies* (pp. 168–187). Oxford, England: Blackwell-Wiley.
Tajfel, H., & Turner, J. (1979). An integrative theory of intergroup conflict. In W. G. Austin & S. Worchel (Eds.), *The social psychology of intergroup relations*, (pp. 33-48). Monterey, CA: Brooks/Cole.
Turner, J. C., Hogg, M. A., Oakes, P. J. Reicher, S. D. & Wetherell, M. S. (1987). *Rediscovering the social group: A Self-Categorization Theory*. Oxford & New York: Blackwell.
Van Dijk, T. (1992) Race, riots and the press. An analysis of editorials in the British press about the 1985 disorders. *Discourse studies*, 43, 229-253
Van Dijk, T. (1984) *Prejudice in discourse*. London: John Benjamins.
Vygotsky, L. S. (1978). Mind in society: The development of higher mental process. Harvard University press
Walton, D. (2007) Media Argumentation: Dialectic, Persuasion and Rhetoric. CUP: Cambridge.
Walton, D. Reed, C., Macagno F. (2008). *Argumentation Schemes*, Cambridge: CUP.
Walton, D., Krabbe E. (1995). Commitment in Dialogue: Basic Concepts of Interpersonal Reasoning, Suny Press
Worchel, S., Grossman, M., & Coutant, D. (1994). Minority influence in the group context: How group factors affect when the minority will be influential. Nelson-Hall Publishers
Zimmerman, M.A., & Rappaport, J. (1988). Citizen participation, perceived control, and psychological empowerment. *American Journal of Community Psychology*, 16, 725-750.

Zuniga, H. Jung N. Valenzuela, S. (2012). Social Media Use for News and Individuals' Social Capital, Civic Engagement and Political Participation. *Journal of Computer-Mediated Communication* 17, 319–336

Chapter 7

Positive and Negative Arguments in Review Systems: An Approach with Arguments

Simone Gabbriellini[1], Francesco Santini[2]

[1] Dipartimento di Economia e Management, Università di Brescia, Italy, simone.gabbriellini@unibs.it
[2] Dipartimento di Matematica e Informatica, Università di Perugia, Italy, francesco.santini@dmi.unipg.it

Abstract. In this work we study arguments in Amazon.com reviews. We manually extract positive (in favour of purchase) and negative (against it) arguments from each review concerning a selected product. Moreover, we link arguments to the rating score and length of reviews. For instance, we show that negative arguments are quite sparse during the first steps of such social review-process, while positive arguments are more equally distributed. In addition, we connect arguments through attacks and we compute Dung's extensions to check whether they capture such evolution through time.

1. Introduction

Recent surveys have reported that 50% of on-line shoppers spend at least ten minutes reading reviews before making a decision about a purchase, and 26% of on-line shoppers read reviews on Amazon prior to making a purchase.[1]

This paper reports a study of how customers use arguments in writing such reviews. We start from a well-acknowledged result in the literature on on-line reviews, i.e. that the more reviews a product gets, the more the rating tend to decrease. Such rating is, in many case, a simple scale 1 to 5, where 1 is a low rating and 5 is the maximum possible rating.

This fact can be explained easily considering that the first customers are more likely to be enthusiast of the product, then as the product gets momentum, more people have a chance to review it and inevitably the average rating tend to stabilise

---

[1] http://www.forbes.com/sites/jeffbercovici/2013/01/25/ how-amazon-should-fix-its-reviews-problem/

on some lower values than 5. Such process, with a few enthusiast innovators followed by a majority that gets convinced by the formers, is a typical pattern in diffusion studies (Rogers, 2003).

However, the level of disagreement in product reviews remains a challenge: does it influence what other customers will do? In particular, what happens, on a lower level, that justifies such diminishing trend in ratings? Since reviewing a product is a communication process, and since we use arguments to communicate our opinions to others (and possibly convince them) (Mercier, 2012), it is evident that late reviews should contain enough negative arguments to explain the negative trend in ratings.

Our present study can be considered as "micro" because we focus on a single product only, even if with a quite large number of reviews (i.e., 253). Unfortunately, due to the lack of tools for the automated extraction of arguments and attacks, we cannot extend our study "in the large" and draw more general considerations. We extracted by hand, for each review about the selected product, both positive and negative arguments expressed, the associated rating (from one to five stars), and the time when the review has been posted. The next step would be to analyse our data in terms of:

- how positive/negative arguments are posted through time.
- how many positive/negative arguments a review has (through time).

In particular, we argue that the reason why average rating tend to decrease as a function of time depends not only on the fact that the number of negative reviews increases, but also on the fact that negative arguments tend to permeate positive reviews, decreasing de facto the average rating of these reviews.

One interesting way to reason with arguments is to adopt a computational argumentation approach. An Abstract Argumentation Framework (AAF), or System, as introduced in a seminal paper by Dung (Dung, 1997), is simply a pair <A,R> consisting of a set A whose elements are called arguments and of a binary relation R on A, called "attack" relation. An abstract argument is not assumed to have any specific structure but, roughly speaking, an argument is anything that may attack or be attacked by another argument. The sets of arguments (or extensions) to be considered are then defined under different semantics, which are related to varying degrees of scepticism or credulousness.

The rest of the paper is structured as follows. Section 2 sets the scene where we settle our work: for instance, we introduce related proposals that aggregate Amazon.com reviews in order to produce a easy-to-understand summary of them. Afterwards, in Section 3 we describe the Amazon.com dataset from where we select our case-study. Section 4 plots how both positive and negative arguments dynamically change through time, zooming inside reviews with a more granular approach. In Sec 6 we draw the Argumentation Framework (Dung, 1997) we can extract from the considered reviews. Finally, Section 7 wraps up the paper and hints direction for future work.

2. Literature Review

Electronic Word of Mouth (e-WoM) is the passing of information from person to person, mediated through any electronic means. Over the years it has gained growing attention from scholars, as more and more customers started sharing their experience online (Anderson, 1998) (Stokes, 2002) (Goldenberg, 2001) (Zhu, 2006) (Chatterjee, 2001). Since e-WoM somewhat influences consumers' decision-making processes, many review systems have been implemented on a number of popular Web 2.0-based e-commerce websites (e.g., Amazon[2] and eBay[3]), product comparison websites (e.g., BizRate[4] and Epinions[5]), and news websites (e.g., MSNBC[6] and SlashDot[7]).

Unlike recommendation systems, which seek to personalise each user's Web experience by exploiting item-to-item and user-to-user correlations, review systems give access to others' opinions as well as an average rating for an item based on the reviews received so far. Two key facts have been assessed so far:

- reporting bias: customers with more extreme opinions have a higher than normal likelihood of reporting their opinion (Anderson, 1998);
- purchasing bias: customers who like a product have a greater chance to buy it and leave a review on the positive side of the spectrum (Chevalier, 2006).

These conditions produce a J-shaped curve of ratings, with extreme ratings and positive ratings being more present. Thus a customer who wants to buy a product is not exposed to a fair and unbiased set of opinions. Scholars have started investigating the relation between reviews, ratings and disagreement among customers (Dellarocas, 2003) (Moe, 2012). In particular, one challenging question is: does the disagreement about the quality of a product in previous reviews influence what new reviewers will post?

A common approach to measure disagreement in reviews is to compute the standard deviation of ratings per product, but more refined indexes are possible (Nagle, 2014). The next step would be to detect correlations among disagreement as a function of time (Dellarocas, 2003) (Nagle, 2014). We aim, however, at modelling a lower level, micro- founded mechanism that could account for how customers' reviewing behaviour evolves over time. We want to analyse reviews not only in terms of rating and length, but also in terms of what really constitutes the review itself, i.e. the arguments used by customers. We aim at explaining disagreement as a consequence of customers' behaviour, not only at describing it as

---

[2] http://www.amazon.com
[3] http://www.ebay.com
[4] http://www.bizrate.com
[5] http://www.epinions.com
[6] http://www.msnbc.com
[7] http://slashdot.org

a correlation among variables (analytical, micro-founded modelling of social phenomena is well detailed in (Manzo, 2013) (Hedstrom, 2005) (Squazzoni, 2012) - for an application in online contexts, see (Gabbriellini, 2014).

However, before automatically reasoning on arguments, we have first to extract them from a text corpora of online reviews. On this side, research is still dawning, even if already promising (Villalba, 2012) (Wyner, 2012). In addition, we would like to mention other approaches that can be used to summarise the bulk of unstructured information (in natural language) provided by customer reviews. The authors of (Hu, 2004) summarise reviews by i) mining product features that have been commented on by customers, ii) identifying opinion sentences in each review and deciding whether each opinion sentence is positive or negative, and 3) summarising the results. Several different techniques are advanced to this, e.g., sentiment classification, frequent and infrequent features identification, or predicting the orientation of opinions (positive or negative). Even if never citing the word "argument", we think (Hu, 2004) is strictly related to argument mining.

3. Dataset

Amazon.com allows users to submit their reviews to the web page of each product, and the reviews can be accessed by all users. Each review consists of the reviewer's name (either the real name or a nickname), several lines of comment, a rating score (ranging from one to five stars), and the time-stamp of the review. All reviews are archived in the system, and the aggregated result, derived by averaging all the received ratings, is reported on the web page of each product. It has been shown that such reviews provide basic ideas about the popularity and dependability of the corresponding items; hence they have a substantial impact on cyber-shoppers' behaviour (Chevalier, 2006). It is well known that the current Amazon.com reviewing system has some noticeable drawbacks (Wang, 2008). For instance, i) the review results have the tendency to be skewed toward high scores, ii) the ageing issue of the reviews is not considered, and iii) it has no means to assess the reviews' helpfulness if the reviews are not evaluated by a sufficiently large number of users.

We retrieved the "Clothing, Shoes and Jeweller" products section of Amazon[8]. The dataset contains approximately 110k products and spans from 1999 to July 2014, for a total of more than one million reviews. The whole dataset contains 143.7 millions reviews.

We summarise here a quick description of such dataset[9]:

- the distribution of reviews per product is highly heterogeneous;

---

[8] Courtesy of Julian McAuley and SNAP project (source: http://snap.stanford.edu/data/web-Amazon.html and https://snap.stanford.edu).
[9] Space constraints prevented to show more detailed results here, but additional plots are available in the form of research notes at http://tinyurl.com/pv5owct.

- the disagreement in ratings tend to rise with the number of reviews until a point after which it starts to decay (interestingly, for some highly reviewed products, disagreement remains high: this means that only for specific products opinions polarise while, on average, reviewers tend to agree)[10];
- more recent reviews tend to get shorter, irrespectively of the number of reviews received, which is pretty much expectable: new reviewers might realise that some of what they wanted to say has already been stated in previous reviews;
- more recent ratings tend to be lower, irrespectively of the number of reviews received.

To sum up, it seems that disagreement in previous reviews does not affect much latest ratings - except for some cases which might correspond to products with polarised opinions. This result has already been found in the literature (Moe, 2012). However, it has also already been challenged by Nagle and Riedl (Nagle, 2014), who found that a higher disagreement among prior reviews does lead to lower ratings. They ascribe their new finding to their more accurate way of measuring disagreement in such J-shaped distributions of ratings. One of the main aims of this work is to understand how it is that new reviews tend to get lower ratings. Our hypothesis is that this phenomenon can be explained if we look at the arguments level, i.e. if we consider, more than aggregate ratings, the dynamic of the arguments used by customers.

Since techniques to mine arguments from a text corpora are yet in an early development stage, we focus on a single product and extracted arguments by hand. We randomly select a product, which happens to be a ballet tutu for kids, and we examine all the 253 reviews that the product received between 2009 and July 2014. From the reviews, we collect a total of 24 positive arguments and 20 negative arguments, whose absolute frequencies are reported in Table 1. We then proceeded to analyse the distribution of arguments as well as some correlations as a function of time, as detailed in the next section.

---

[10] Polarisation only on specific issues has already been observed in many off-line con- texts, see (Baldassarri, 2007).

ID	Positive arguments	#App.	ID	Negative arguments	#App.
A	the kid loved it	78	a	it has a bad quality	18
B	it fits well	65	b	it is not sewed properly	17
C	it has a good quality/price ratio	52	c	it does not fit	12
D	it has a good quality	44	d	it is not full	11
E	it is durable	31	e	it is not as advertised	8
F	it is shipped fast	25	f	it is not durable	7
G	the kid looks adorable	23	g	it has a bad customer service	4
H	it has a good price	21	h	it is shipped slow	3
I	it has great colors	21	i	it smells chemically	3
J	it is full	18	j	you can see through it	3
K	it did its job	11	k	it cannot be used in real dance class	2
L	it is good for playing	11	l	it has a bad quality/price ratio	2
M	it is as advertised	9	m	it has a bad envelope	1
N	it can be used in real dance classes	7	n	it has a bad waistband	1
O	it is aesthetically appealing	7	o	it has bad colours	1
P	it has a good envelope	2	p	it has high shipping rates	1
Q	it is a great first tutu	2	q	it has no cleaning instructions	1
R	it is easier than build your own	2	r	it is not lined	1
S	it is sewed properly	2	s	it never arrived	1
T	it has a good customer service	1	t	it was damaged	1
U	it is secure	1			
V	it is simple but elegant	1			
W	you can customize it	1			
X	you cannot see through it	1			

Table 1. Positive and negative arguments, with their number of appearances in reviews between 2009 and July 2014.

4. Analysis

In Figure 1, the first plot on the left shows the monthly absolute frequencies of positive arguments in the specified time range. As it is easy to see, the number of positive arguments increases as time goes by, which can be a consequence of a success in sales: more happy consumers are reviewing the product. At the same time, the first plot on the right shows a similar trend for negative arguments, which is a signal that, as more customers purchase the product, some of them are not satisfied with it. According to what we expect from the literature, the higher volume of positive arguments is a consequence of the J-shaped curve in ratings, i.e., a consequence of reporting and selection biases. What is interesting to note though is that the average review rating tend to decrease with time, as shown by the second row of plots. This holds both for reviews containing positive arguments as well for those containing negative arguments. In particular, the second plot on the right shows that, starting from 2012, negative arguments start to infiltrate "positive" reviews, that is reviews with a rating of 3 and above.

Finally, the last two plots in Figure 1 show that the average length of reviews decreases as time passes both for reviews with positive arguments than for ones with negative arguments. However, such a decrease is much more steep for negative reviews than for positive ones.

In Figure 2 we can observe the distribution of positive and negative arguments.[11] Regarding positive arguments, we cannot exclude a power-law model for the distribution tail with x-min = 18 and α = 2.56 (pvalue = 0.54).[12] We also tested a log-normal model with x-min = 9, μ = 3.01 and σ = 0.81 (pvalue = 0.68). We then searched a common x-min value to compare the two fitted distributions: for x−min = 4, both the log-normal (μ = 3.03 and σ = 0.78) and the power-law (α = 1.55) models still cannot be ruled out, with p–value = 0.57 and pvalue = 0.54 respectively. However, a comparison between the two leads to a two-sided pvalue = 0.001, which implies that one model is closer to the true distribution - in this case, the log-normal model performs better. For negative arguments, we replicated the distribution fitting: for xmin = 2, a power law model cannot be rule out (α = 1.78 and p-value = 0.22) as well as a log-normal model (μ = 1.48 and σ = 0.96, pvalue = 0.32). Again, after comparing the fitted distributions, we cannot rule out the hypotheses that both distributions are equally far from the true distribution (two-sided pvalue = 0.49). In this case, too few data are present to make a wise choice.

The plots in Figure 3 show the cumulative frequencies of each single arguments as a function of time. The frequencies are calculate over all the arguments (repeated or new) across all reviews, so to give an idea of how much a single argument represents customers' opinion. Among the positive arguments (plot on the left), there are four arguments that represent, taken together, almost 44% of customers' opinions. These arguments are:

- good because the kid loved it
- good because it fits well
- good because it has a good quality/price ratio
- good because it has a good quality

---

[11] We used the R poweRlaw package for heavy tailed distributions developed by Colin Gillespie. The logic of the fitting procedures is available in [17]

[12] We used the relatively conservative choice that the power law is ruled out if p value 0.1, as in [11].

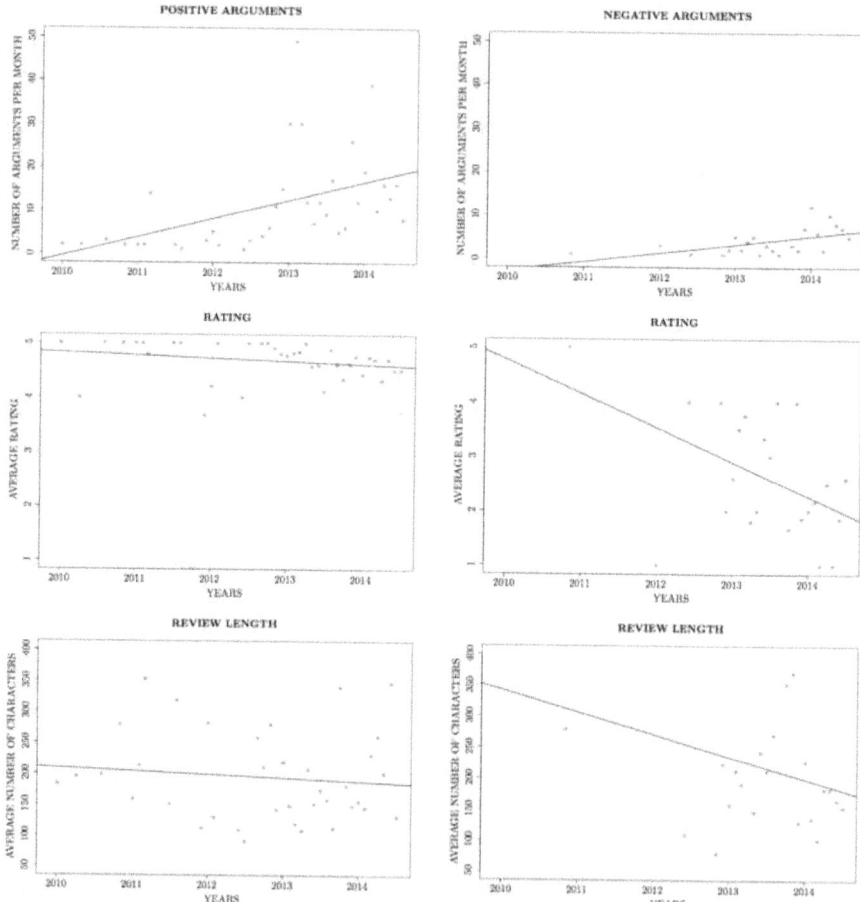

Figure 1. Argument trends: row1, the absolute frequency of arguments, row2 the average rating of reviews, row3 the average review length per month

Positive and negative arguments in review systems

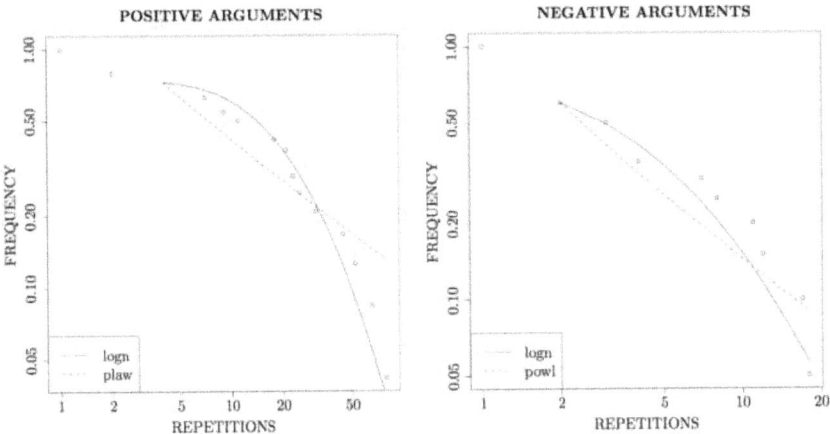

Figure 2. Arguments distribution: probability of observing an argument repeated x times.

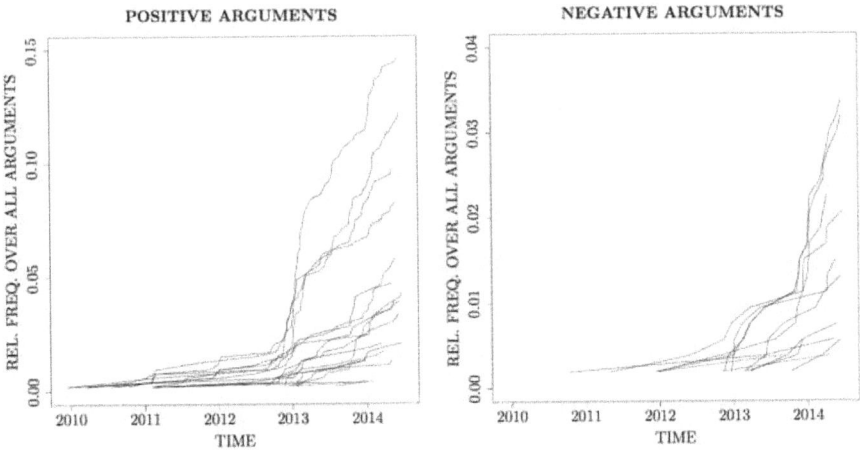

Figure 3. Relative cumulated frequencies for each positive and negative argument.

Negative arguments represent, all together, less than 20% of opinions (plot on the right). As expected, they are less repeated and less frequent than positive ones. Among these arguments, two of them have the higher impact:
- bad because it has a bad quality
- bad because it is not sewed properly

125

We have a clear situation where the pros and cons of this product are clearly stated in the arguments: not surprisingly, the overall quality is the main reasons why customers will consider the product as a good or bad deal. Even among unsatisfied customers, this product is not considered expensive, but quality still is an issue for most of them.

The plots in Figure 4 show the cumulative frequencies and the rate at which new arguments are added as a function of time. In the left plot, it is interesting to note that, despite the difference in volume (positive arguments are more cited than negative ones, as detailed in previous plots), the cumulative frequencies at which positive and negative arguments are added are almost identical. Positive arguments start earlier than negative ones, consistently with the fact that enthusiast customers are the first that review the product. At the same time, it's interesting to note that no new positive argument is added in the 2011-2013 interval, while some negative ones arise in the reviews. Since 2013, positive and negative arguments follows a pretty similar trajectory. However, as can be noted in the second plot on the right, new arguments are not added at the same pace. If we consider the total amount of arguments added, positive ones are repeated more often than negatives, and the rate at which a new positive argument is added is considerably lower than its counterpart. This information shed a light on customers' behaviour: dissatisfied customers tend to post new reasons why they dislike the product, more than just repeating what other dissatisfied customers have already said.

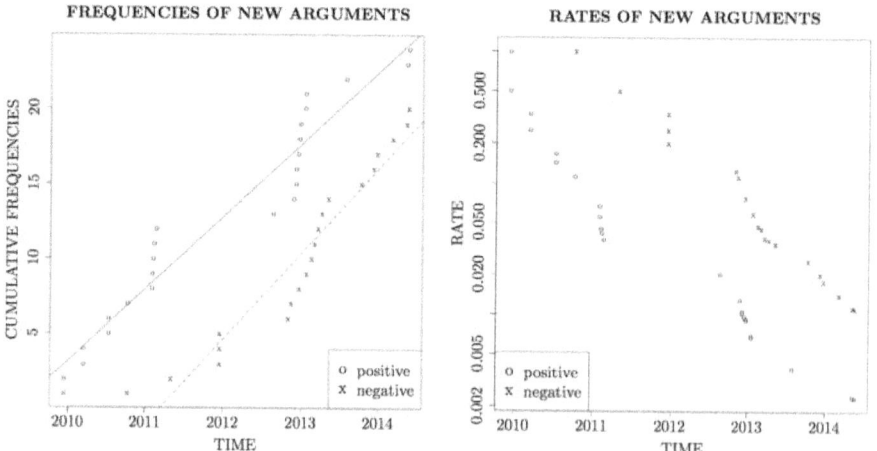

Figure 4. Left plot: cumulative frequencies of new positive and negative arguments per month. Right plot: rate of new positive and negative arguments over total arguments per month.

5. Abstract Argumentation and Our Running Example

In this section we start by briefly summarising some background information related to classical Abstract Argumentation Frameworks (AAFs) (Dung, 1997).

Definition 1 (AAF). An Abstract Argumentation Framework (AAF) is a pair F = <A,R> of a set A of arguments and a binary relation R ⊆ A × A, called the attack relation. ∀a,b ∈ A, a R b (or, a → b) means that a attacks b. An AAF may be represented by a directed graph whose nodes are arguments and edges represent the attack relation. A set of arguments S ⊆ A attacks an argument a, i.e., S → a, if a is attacked by an argument of S, i.e., ∃b ∈ S. b → a.

In the following of this section, we collect the information manually extracted in the previous section with the purpose to draw a corresponding AAF.

In addition to arguments (see Section 5), even attacks among them have been "manually" extracted, with the purpose to represent all the knowledge as an AAF. For some couples of arguments, this has been very easy: some positive arguments are the exact negation of what stated in the relative negative argument (and vice-versa). For instance, looking at Table 1, the tutu has a good quality (D) and the tutu has a bad quality (a), or the tutu fits well (B) and the tutu does not fit (c). For the sake of completeness, such easy-to-detect (bidirectional) attacks are {B ↔ c, C ↔ l, D ↔ a, E ↔ f, F ↔ h, I ↔ o, J ↔ d, M ↔ e, N ↔ k, P ↔ m, S ↔ b, T ↔ g, X ↔ j}. Furthermore, we have identified some other unidirectional and bidirectional attacks; the complete list of arguments is visually reported with the graphical representation of the whole AAF, in Figure 5. Note that we also have two unidirectional attacks between two positive arguments (Q → N and V → J), and one bidirectional attack between two negative arguments (s ↔ h). Some of the reported attacks need the full sentences (or even the whole reviews) related to the extracted arguments, in order to be comprehended at full. Even if sometimes connections are hidden at first sight, we tried to be as more linear as possible, thus avoiding too much subtle criticisms.

The complete AAF is represented in Figure 5: circles represent positive arguments, while diamonds represent negative ones. We have clustered most of the arguments into four main subsets: a) product quality, b) product appearance, c) shipping-related information, and d) price-related information. Four arguments do not belong to any of this subsets (see Figure 5).

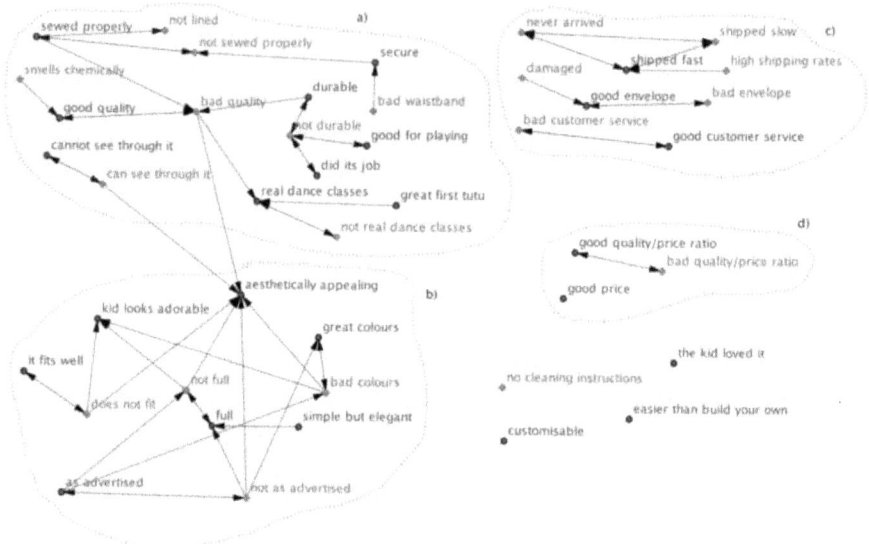

Figure 5. The final and complete AAF extracted from 253 reviews.

6. Conclusion and Future Work

In this paper we have proposed a first case study on how to use Abstract Argumentation to understand if using arguments can improve our knowledge about social trends in product reviews. More in particular, we "enter" into an Amazon.com review and we achieve a more granular view of it by considering the different arguments expressed in each of the 253 reviews about the selected product (a ballerina tutu). What we observe is that the frequency of negative arguments (against purchasing the tutu) increases after some time, while the distribution of positive arguments (in favour of purchasing the tutu) is more balanced between the considered period. Moreover, while positive arguments are always associated with high ratings (i.e., 4 or 5), negative arguments are associated with low (as expected) but also high ratings. In addition, negative arguments are more frequently associated with shorter reviews, while enthusiasts tend to be less concise. To summarise, the aim is to "explode" reviews into arguments and then try to understand how the behaviour of reviewers changes through time, from the arguments point of view.

Our present study can be considered as "micro" because we focus on a single product only, even if with a quite large number of reviews (i.e., 253). Unfortunately, due to the lack of tools dedicated to the automated extraction of AAFs, we cannot extend our study "in the large" and draw more general considerations. Argument mining is a relatively new challenge in corpus-based discourse analysis, and it involves the automatic identification of argumentative structures within a document (e.g., the premises), as well as the relationships between pairs of arguments. In our manual extraction we noticed that, if extracting abstract arguments from natural language is not easy, the process of recognising attacks is even more challenging, due to subtle criticisms and, in general, ambiguities of natural languages.

In the future, we hope to widen our investigation by taking advantage of the (expected) improvement of preliminary mining-techniques proposed in (Villalba, 2012) and (Wyner, 2012), which are more related to product reviews, as in our case study. In addition, we plan to understand if tolerating a given low amount of inconsistency (i.e., attacks) in extensions (Bistarelli, 2010) can help softening the impact of weak arguments (i.e., rarely repeated ones). Due to the possible partitioning of arguments into clusters related to different aspects of a product (e.g., either its quality or appearance), we also intend to apply coalition-oriented semantics, as proposed in (Bistarelli, 2013).

Following (Gabbriellini, 2014), we also plan to implement an Agent-Based Model with Argumentative Agents to explore the possible mechanisms, from a user's perspective, that give raise to such trends and correlations among positive and negative arguments.

References

Anderson, E. (1998). Customer satisfaction and word of mouth. Journal of Service Research, 1-7.
Baldassarri, D., & Bearman, P. (2007). Dynamics of political polarization. American Sociological Review.
Bistarelli, S., & Santini, F. (2010). A common computational framework for semiring-based argumentation systems. European Conference on Artificial Intelligence. FAIA, 131-136.
Bistarelli, S., & Santini, F. (2013). Coalitions of arguments: An approach with constraint programming. Fundamenta Informaticae, 383-401.
Chatterjee, P. (2001). Chatterjee, P. . ACR 2001 Proceedings.
Chevalier, J., & Mayzlin, D. (2006). The effect of word of mouth on sales: Online book reviews.
Dellarocas, C. (2003). The digitization of word of mouth: promise and challenges of online feedback mechanisms. Management Science .
Dung, P. (1997). On the acceptability of arguments and its fundamental role in non-monotonic reasoning, logic programming and n-person games. Artificial Intelligence, 321–357.
Gabbriellini, S. (2014). The evolution of online forums as communication networks: An agent-based model. Revue Francaise de Sociologie , 805-826.

Gabbriellini, S., & Torroni, P. (2014). A new framework for abms based on argumentative reasoning. Advances in Social Simulation, 25-36.

Goldenberg, J., Libai, B., & Muller, E. (2001). Talk of the network: a complex systems look at the underlying process of word-of-mouth. Marketing Letters.

Hedstrom, P. (2005). Dissectin the Social: on the Principles of Analytical Sociology. Cambridge University Press.

Hu, M., & Liu, B. (2004). Mining and summarizing customer reviews. SIGKDD International Conference on Knowledge Discovery and Data Mining, ACM, 168-177.

Manzo, G. (2013). Educational choices and social interactions: A formal model and a computational test. Comparative Social Research, 47-100.

Mercier, H., & Sperger, D. (2012). Why do humans reason? Arguments for an argumentative theory. Behavioral and Brain Sciences.

Moe, W., & Schweidel, D. (2012). Online product opinions: Incidence, evaluation, and evolution. Marketing Science .

Nagle, F., & Riedl, C. (2014). Online word of mouth and product quality disagreement. Academy of Management .

Rogers, E. (2003). Diffusion of Innovations. Simone & Schuster.

Squazzoni, F. (2012). Agent-Based Computational Sociology. Wiley.

Stokes, D. L. (2002). Taking control of word of mouth marketing: the case of an entrepreneurial hotelier. Journal of Small Business and Enterprise Development.

Villalba, M., & Saint-Dizier, P. (2012). A framework to extract arguments in opinion texts. IJCINI, 62-87.

Wang, B. Z. (2008). Improving the amazon review system by exploiting the credibility and time-decay of public reviews. Web Intelligence and Intelli- gent Agent Technology, 123-126.

Wyner, A., Schneider, J., Atkinson, K., & T.J.M., B.-C. (2012). Semi-automated argumentative analysis of online product reviews. Computational Models of Argument - Proceedings of COMMA 2012. FAIA, 43-50.

Zhu, F., & Zhang, X. (2006). The influence of online consumer reviews on the demand for experience goods: The case of video games. Proceedings of the International Conference on Information Systems, 25.

Chapter 8

Hypotheses of Analysis on the Stylistics of Arguments: a Case Study from Trip Advisor

Laura Bonelli

Università "La Sapienza" and ISTC-CNR; Rome, Italy , laura.bonelli@istc-cnr.it

Abstract. User generated content and cWOM (Electronic Word Of Mouth) on travel related information are increasingly widespread online practices from which travelers take advantage of (Gretzel & Yoo, 2008). Travelers' reviews are based on persuasive arguments aimed at convincing other travelers of the truthfulness of their experience and of the validity of their opinions. Starting from the idea that semantic and expressive levels of argumentation should not be seen as separated instances, this chapter proposes an analytic framework of online sentiment analysis, which mainly focuses on the users' stylistic and rhetoric choices in a corpus of 100 Trip Advisor reviews of several hotels based in London, UK. The users' communicative behavior is depicted by means of Caffi and Janney (1994)'s emotive devices. Results show that most devices work as reinforcements of the users' inferred evaluative standpoints. The proposed analytic framework may also posit itself as complementary to more traditional frameworks of argumentative analysis.

1. Introduction

Argumentation is aimed at providing and assessing reasons in communicative exchanges: it has both an interpersonal and an intrapersonal function, depending on contexts and goals. However, as speakers of this world, we face a series of daily challenges when we try to prove our points: we need to properly highlight our reasons at the right time, with the most felicitous attitudes towards our interlocutors, with appropriate words and tone of voices, and with the most persuasive word order.

These types of micro-stylistic, strategic choices are a relevant part of the meaning of our messages and they shouldn't be neglected in the analysis of arguments: how we decide to express what we mean is exactly what makes our arguments more or less persuasive in our contexts of interaction. Whereas face-to-face interactions may lead us to often smooth the communicative management and likely reduce the level of assertivity and even of certainty of our arguments, online exchanges may, on the contrary, let us feel more protected in terms of face-saving needs, and lead us to use more direct and straight-to-the-point strategies, especially if an online community with relatable needs is our main audience, and being of any help to them is among our primary goals.

This often happens to be the case of Trip Advisor, an American travel website company made popular by its early employment of user generated content: on its portal, among other services, users are allowed to share their experiences by writing reviews of the hotels, restaurants, vacation houses or touristic attractions they've seen or visited, giving feedbacks to managers, expressing their feelings, and helping other travelers at the same time. The service is free of charge, easily accessible, largely popular and remarkably influential: eWOM ("Electronic Word Of Mouth") appears to be more actively requested and even more credible especially in those contexts where online users need to make decisions on something unfamiliar, as it often happen with tourism related choices (Gretzel & Yoo 2008).

Trip Advisor reviews assess and construct the reputation of each item with different parameters, the main one resulting in ratings on a five point scale that go from "terrible" to "excellent". The research I carried tries to shed light on the communicative strategies that different users employed to negatively (1-2 points), positively (4-5 points) or averagely (3 points) review a series of hotels in London, (UK) on tripadvisor.com, and how these strategies tend to highlight specific attitudes aimed at making their arguments more persuasive. In order to do so, I analyzed the users' standpoints and arguments using Caffi & Janney (1994) and Caffi (2001; 2007)'s linguistic pragmatics approach, an approach that puts linguistic style in the foreground. In particular, I carried my analysis using these authors' six markers of evaluation, proximity, specificity, evidentiality, volitionality and quantity, in order to depict the users' persuasive strategies in terms of the micro-stylistic choices they employed to express their arguments. Results show three different recurring tendencies of markedness in the negative, average, and positive reviews and a set of possibly interesting correlations between the markers. But before heading to data analyses and results, I will first try to give reason for my methodological choice, which can seem quite peculiar for argumentation researchers.

2. Rhetorics and Pragmatics: Common Goals and Merging Points

Before even considering the various argumentation theories and approaches, a connection between rhetorics and linguistic pragmatics might as well look incautious at first sight, especially if we think of – *mutatis mutandis* – the nature of both disciplines, which is not always internally cohesive. However, this connection still finds its *raison d'être* in multiple common viewpoints and goals which the two disciplines value and share. Rhetorics shares a series of crucial similarities with pragmatics: "both trigger a *surplus de sense*, both presuppose shared knowledge on the speaker's and hearer's parts" and, most importantly, "both rely on the hearer's cooperation and willingness to undertake the inferential steps necessary to give utterances intended meanings beyond their literal ones" (Caffi & Janney 1994: 330).

Moreover, the two disciplines share the expression and the negotiation of emotionality between speaker and hearer (Venier, 2008: 97) and in the construction of that necessary knowledge which we, as speakers of this world, employ in order to create and share efficacious and persuasive messages (see also Leech, 1980, 1983; Caffi, 1989, 1990). From a historical point of view, illustrious ancestors of these reflections can be found in Aristotelian rhetorics (especially in its presentation of how different types of argumentation suit different types of audiences), in Pseudo-Longinus' idea of πάθος and in Perelman and Olbrechts-Tyteca (1958)'s Traité de l'argumentation, whose 'nouvelle rhétorique' describes the complex emotive strategies that orators must enact in order to completely adapt to their audience, create consensus, reinforce their expressive strategies, and stimulate the emotive participation of their public.

Rhetorics and linguistic pragmatics share an essentially intersubjective orientation. They also both see human communication as a process that is fully anchored in metapragmatic awareness (Verschueren 2000; Caffi 2006; Rigotti & Greco Morasso 2010), which refers to that type of understanding that allows an assessment of the discourse's appropriateness both from a formal and from a conceptual point of view. The two disciplines both stress their attention on the idea of a communicative willingness that can fill discourses with the strategic signaling of specific intentions (Hübler, 1987; Mortara Garavelli 1988). They are both oriented to polytropy and "recipient design" (Perelman & Olbrechts-Tyteca, 1958; Sacks, Schegloff & Jefferson, 1974; Giles, 1979; Giles, Coupland & Coupland, 1991; Caffi 2001) and they both deal with the practice of inferring and expressing opinions (Caffi 2006; Rigotti & Greco Morasso 2010).

All of these common aspects lead to an idea of both epistemic and affective commitment in the way we express our points of view, in the way we stress and mark precise words, in the way we organize the topics in our utterances, or in the way we, on the contrary, measure, hedge and downgrade the tone or the relevance of our topics, thinking of our interlocutors' attention and attunement as a goal and as a priority. The way we express this commitment is by means of precise stylistic choices, which have a considerable impact on the way our arguments are understood and accepted.

Current argumentative approaches are far from neglecting the communicative organization of their units of analysis. However, despite their research origins were strongly intermingled with rhetorics, they specialized in such directions that backgrounded most stylistic issues in the analysis of arguments, and at times these are simply not considered semantically relevant.

Linguistic pragmatics exhibits a fundamental theoretical fluidity and its intersection with many disciplines, rather than a weakness, could be valued as a strength. This is especially true from the viewpoint of a systemic approach that also takes semantic dimensions into consideration, without disregarding or denying attention to rhetoric and stylistic micro-phenomena (Caffi 2001: 146).

If an actual merging point on these phenomena could be expected, it has to be imagined on the semantics-pragmatics interface (*e.g.*: Rigotti & Rocci 2006 on theme-rheme; Caffi 2001 on topical shields).

3. Not a Matter of Beauty: Relevant Issues in Historical Stylistics

Communicative actors are highly interested in the most persuasive ways in which they can convey their opinions' truthfulness, or at least their epistemic or affective commitment to them. As I hinted at above, such a viewpoint seems to imply that communicative style (1) would better be conceptualized in such a way that no a priori separation between semantic and expressive levels is contemplated and (2) it can be understood as a powerful tool by means of which speakers actively co-construct their path to seek attention from and understand each other (see Arndt & Janney, 1987; Caffi, 1990; Caffi & Janney, 1994; Caffi, 2001 on meaning co-construction).

The point delineated above in (1) has not always been equally shared by stylistics scholars. In her work on "linguostylistics", Akhmanova (1976: 3) claims that "linguistic style is that part of language which is used to impart to the message certain expressive-evaluative-emotional overtones, […], and expressive-evaluative-emotional overtones are superimposed on the main semiotic content". In his "pragmastylistics", Hickey similarly thinks that (1989: 2) "[…] style is what distinguishes one utterance from another when both denote the same, or approximately the same, thought: it is the aspect of an expression or text which contrasts with another when both communicate 'the same' content". Taylor (1980: 3), talking about verbal style, claims that "it has had to be defined as a peripheral aspect of communication, distinct from, but dependent upon the principal function: the communication of meaning". On the same wavelength, Wales (2006: 1046) writes: "[…] in any language there is always more than one way of writing or speaking the same message, but with a different connotation or effect as a result".

Such viewpoints tend to evaluate rhetorical and stylistic choices as additional, unnecessary embellishments of a message's content, whereas the content itself is seen as stable and impervious. A clear-cut separation between the stylistic aspects of a message and its meaning, however, does not consider that each choice of formulation corresponds to a different discursive strategy, which conveys specific meanings and intentions. What Akhmanova calls "expressive-evaluative-emotional overtone" is actually something that marks a specific psychological attitude on the expressive level, whose function is to strategically compose meanings that cannot be entirely conveyed without certain specific, localized verbal and non-verbal choices. A different point of view in this sense can be found in Bally's [1909], (1921: 16) definition of stylistics, which mostly takes the speakers into account: "*La stylistique étudie donc les faits d'expression du langage organisé au point de vue de leur contenu affectif, c'est-à-dire l'expression des faits de la sensibilité par le langage et l'actions des faits de langage sur la sensibilité*".

As I mentioned above in (2), communicative style could also be understood as a medium by means of which communicative partners create a cognitive and emotive interdependence as they try to persuade and understand each other. This interdependence is expressed and interpreted within a complex system of expectations (Miceli and Castelfranchi 2014). Part of the significance of each expressive choice is given by their degree of contrast with other hypothetical expressive choices that can be considered more usual or appropriate in a given context. In this sense, stylistics – as well as linguistic pragmatics – deals with the idea of norm. Delbouille (1960: 94) notices how conversational or literary style itself can be understood as a deviation from the type of norms of language use we can find in neutral or in more decontextualized expressions. In his work on French stylistics, Bally [1909: 28-9] sees the idea of norm as part of the stylistician's methodology: "*Un principe important de notre méthode, c'est l'établissement, par abstraction, de certains modes d'expression idéaux et normaux; ils n'existent nulle part à l'état pur dans le langage, mais ils n'en deviennent pas moins des réalités tangibles*". According to Riffaterre (1960)'s stylistics, a stylistic marker is any linguistic element that stands out from its context in such a way that attention and reactions are instantly dragged and elicited in the reader. Marked expressive choices are instantly perceived as relevant for the very fact that they are less expected. Once again, in this sense it is quite hard to think of style as a mere decorative embellishment.

4. Linguistic Markedness and Caffi & Janney (1994)'s Emotive Devices

Bally and Riffaterre's considerations on style as intrinsically affective, semantically relevant, and strategically contrastive with norms and context bound expectations bring attention on what kind of linguistic and semiotic devices can be used as analytic tools in the case we want to see how style works in argumentation.

Communicative outcomes are overflowing with polyfunctional signals or markers (Hölker, 1988) which indicate the quality of the self-presentation enacted by the speakers and the quality of their cooperation with their interlocutors on different levels (*e.g.*: prosodic, morpho-syntactic, stylistic and rhetorical levels). The idea of strategic *markedness* of discursive contents and modalities (Hübler, 1987) has a long tradition in semiotic studies, as well as in social sciences (see for instance Abercrombie, 1967). The signaling of speech markers is a metacommunicative activity through which the speakers can negotiate needs, request and express information, and regulate personal attitudes (Caffi, 2001: 26).

From the point of view I am assuming in this work, it is important to identify a comprehensive operational category of markers able to detect and integrate the speakers' attitudes and the stylistic modality in which the communicative content is expressed. Caffi and Janney's emotive devices are a direct attempt of gathering

Giles, Scherer and Taylor (1979)'s speech markers and Gumperz (1982)'s contextualization cues in a unique polyfunctional type of analytic tools.

The authors identified six different *emotive devices* based on the three most recurrent psychological dimensions of affect in the history of psychology – *evaluation*, *potency* and *activity* – (Osgood et al., 1957), and on the most widespread linguistic categories up to the early 90's.

Rather than focusing solely on the propositional content of the conversational units of analysis (thus investigating communication not exclusively on its semantic and lexical levels), Caffi and Janney (1994: 354) preferred to dedicate special attention to the communicative phenomena that could highlight a certain global affective tonality of the conversation, and they did so by systematically organizing the different types of rhetorical, stylistic and possibly prosodic and paralinguistic choices that the speakers use in order to strategically produce different evocative effects connected with the kind of stance they display.

The devices they proposed are:

1. *evaluation devices* (polarity: positive/negative), which include all the verbal and non-verbal choices used to assess the speaking partner or the discursive content and context (*e.g.:* friendly or hostile tones of voice, modal adverbs, adjectives, vocatives, diminutives, lexical or stylistic choices conveying a positive or a negative attitude). According to the authors, these choices can be interpreted as indexes of pleasure or displeasure, agreement or disagreement, sympathy or antipathy.
2. *proximity devices* (polarity: close/far), which include all the verbal and non-verbal choices that can modify the metaphorical distances between the speakers and their conversational topics, between the speakers and the spatial and/or temporal objects belonging to their speaking context, or among the speakers themselves. Proximity is intended as a subjective dimension emotively experienced by the speakers and aimed at the shortening (or at the widening) of their own perceived distances, including the communicative ways of approach or withdrawal towards specific objects of appraisal.
3. *specificity devices* (polarity: clear/vague), which include all the lexical choices, conversational techniques, and those organizational patterns in the utterance that can express a variation in the level of clarity and accuracy regarding objects and states of affair, the interlocutor, and the conversation itself. Examples are direct or indirect vocatives, definite articles and pronouns *vs* indefinites, generic references to the whole *vs* specific references to parts of a whole (*e.g.:* "Lunch was great" / "The salad dressing was great"), explicit subjects *vs* generic subjects (*e.g.:* "I think that" / "One thinks that").
4. *evidentiality devices* (polarity: confident/doubtful), which include all the linguistic strategies that can regulate the speaker's subscription to

the correctness and credibility of what she intends to speak of. From the point of view of an emotive approach to conversation, the most interesting feature of these devices is their ability to convey the speaker's level of confidence or insecurity towards specific topics and interlocutors (1994: 357). Examples are strategic uses of modal verbs (*e.g.*: "It's correct" / "It might be correct"), the degree of explicitness of an intention (*e.g.*: "I'm coming tomorrow" / "I might be coming tomorrow"), other sorts of parentheticals, modal adverbs, hedges (Brown & Levinson, 1987; Lakoff, 1972), verbal forms of epistemic commitment (Schiffrin, 1987; Lyons, 1977), verbal forms of self-identification with the conversational topic (Tannen, 1989), and more generally all the prosodic and non-verbal choices that can express a major or minor level of intended clearness.

5. *volitionality devices* (polarity: assertive/non-assertive), which include all the linguistic and conversational strategies that can give the conversational agents an active or a passive role. Examples are, again, strategic uses of modal verbs in requests (*e.g.*: "Would you mind passing the salt?" / "Can you pass the salt" / "Give me the salt"), or active *vs* passive verbal forms in regards to expressing opinions (*e.g.*: "I thought that" / "It was claimed that"). The research on volitionality phenomena is central in studies of Western politeness (*inter alia*: Brown & Levinson, 1987; Blum-Kulka, 1987; see Locher & Graham, 2010 for a recent overview).

6. *quantity devices* (polarity: more intense/less intense), which include all the lexical, prosodic and sometimes kinesic choices aimed at enhancing or reducing the level of conversational intensity (Volek, 1987; Labov, 1984). Eterogeneous examples are unexpected prosodic stress (*e.g.*: "Don't do that" / "DON'T do that!"), emphatic adjectives (*e.g.*: "It was a good experience" / "It was an awesome experience"), adverbs (*e.g.*: "It was quite / definitely fun"), and various rhetorical strategies of repetition (*e.g.*: "I'm happy, really happy we have met").

The emotive devices of evaluation, specificity and evidentiality often seem to foreground the speaker-content relationship and to background the speaker-interlocutor relationship, while the devices of volitionality appear to be crucial in the speaker-interlocutor relationship, but less important in the speaker-content relationship. When the focus of the communicative act is the interlocutor, preferred choices are rhetorical and stylistic strategies aimed at expressing the willingness to maintain the interlocutor's approval, displays of respect (*i.e.*: low levels of assertiveness, recurring positive evaluations, high levels of vagueness and politely doubtful choices) and face-saving strategies (Brown & Levinson, [1978], 1987; Goffman, 1971, among others). When the focus of the communicative act is the speaker herself instead, preferred choices are self-disclosures and choices related to the speaker's own attitudes and desires, primarily marked by devices of evaluation and proximity, and enhanced by devices of quantity. Finally, when the focus of the communicative act is the conversational content, devices of (2.4) and generally the order in which the elements appear in each utterance are especially central in the expression of relevance and proximity to specific objects and states of affair.

5. Data and Choice of Methods

The corpus I analyzed is made up of 100 Trip Advisor reviews on different hotels based in London, (UK), of which 34 were negative reviews (from 1 to 2 points), 33 were average reviews (3 points), and 33 were positive reviews (from 4 to 5 points). The average length of the reviews was of 153,76 words each. No socio-demographic data on the reviewers were made available.

I carried a preliminary, word-by word analysis of each review focused on:

- the reviewers' lexical, morphological, syntactic and rhetoric choices;
- paralinguistic strategies (*e.g.:* the strategic use of punctuation or of capital letters with emphatic purposes);
- semantic and metacomunicative strategies (*e.g.:* fictionalization or irony in possible segments of storytelling);
- stylistic strategies, specifically intrastylistic strategies concerning strategic style shifts and the degrees of directness and indirectness of their evaluations;
- types of illocutionary acts.

In a second moment, I used Caffi & Janney's emotive devices of evaluation, proximity, specificity, evidentiality, volitionality and quantity to treat and quantify each of the qualitative communicative behaviors the users adopted in their reviews to describe their experiences. The devices are of multimodal nature and they could specifically detect and describe the users' linguistic and paralinguistic strategies previously taken into account (see Figure 1). Specifically, evaluation devices were treated as evaluative standpoints, whereas devices of specificity, evidentiality, volitionality and quantity, which respectively marked more or less specific, certain, assertive, and intense expressions, were hypothesized as explicit supports of those standpoints. Evaluations mainly referred to the hotels' locations, facilities, price, cleanliness, and staff's qualitative features such as competence and friendliness.

N6

"Should be closed down"

Should have wrote this review straight away but never got round to it. This B & B was the most disgusting place we had ever booked. ¶
On arrival we were told that our room was not ready as the cleaners were still in there. When we got our keys for the room we found that the room had not been touched from the previous people, the bed still had the previous peoples bedding on. The room was awfull. ¶
We complained and said that we were not happy and demanded our money back. To cut a long story short, the owner came (after demanding that he met us for a refund) and he took us, as what he described his best room. I was shocked at what I saw as when I started to check the bed (as the previous mattress in the rooms before were stained as what had to be urine) The owner / manager said that I did not need to check the mattress as it was clean, no it was not as at least half the bed was stained not only with urine it had blood staines, the manager did not think this was a problem. I walked out and demanded my money back. After complaining to the people we booked through and the manager we did after a struggle get all our money back. And yes it was us that reported you to Environmental Health. Disgusting place, hence we never stayed there and booked elswhere. ¶
If you have booked this place, PLEASE check the room first before any payment is made. ¶

Turn	Type of outcome (contextual and co-textual description)	Illocutionary acts	Syntactic, morphological and lexical choices	Rhetorical and stylistic choices	Metacomunicative and metadiscoursive acts	Mitigating devices and boosters
0	The user writes a title for his review	Representative and directive acts	Strategic use of conditional verbs		Choice of a title	
1a	The user provides a first, negative assessment of the hotel	Representative acts	"Should... it" as reinforcing premise of "This B&B... booked"		Metacommunicative assessment ("Should... it")	Boosters: "straight away", "ever"
1b	The user describes his arrival	Representative acts			Micro-sequence of storytelling ("On arrival... bedding on")	
1c	The user complains about his/her experience	Representative and expressive acts		Parentheses. Sarcasm. Style shifts. Style shift, enallage	Storytelling sequence ("We complained... elswhere"): complaint story	Boosters: "And yes", "not only", "at least half of the bed".

Figure 1. Extract of one of the preliminary word-by-word analyses. I hypothesized a micro-turn structure for each review and considered the beginning of a new micro-turn every time the user explicitly started a new paragraph.

In this type of corpus, evidence was too little to consider proximity devices in all of their sub-senses, as outlined in Caffi & Janney (1994)'s extended hypothesis. Most instances of storytelling in the reviews were described with past tenses, making the detection of temporal distance and proximity too fuzzy. Proximity devices were thus detected only in their spatial sense (*e.g.*: in expressions that remarked the hotel's convenient or inconvenient location) and in their social sense (*e.g.*: in expressions that highlighted more or less familiarity with the staff, for instance by calling them with their names or even nicknames instead of generally referring to them). Despite this limitation, proximity devices were as well hypothesized as explicit supports of the reviewers' evaluative standpoints (*e.g.*: unfavorable locations were likely associated with negative evaluations, whereas familiarity with staff or convenient locations were likely associated with positive evaluations).

In a third and final moment, I measured the average distribution of the devices in the negative, average and positive datasets. The units of analysis being reviews, devices of positive and negative evaluation were predominant. Bearing this in mind, I kept track of whether and how proximity, specificity, evidentiality, volitionality and quantity devices were used by the users as supportive strategies to reinforce their evaluative standpoints by means of linear correlations.

6. Results

Positive, average and negative reviews showed different patterns of markedness. On average, negative reviews exhibited relatively higher degrees of specificity (14% of the total devices) and moderate degrees of intensity (15% of the total devices). Positive reviews, on the other hand, tended to be slightly more vague (7% of the total devices) and more intense (19% of the total devices). Few strategies of mitigation have been generally recorded. Even though they always clearly appeared as stylistic strategies aimed at intensifying the evaluative standpoints, the displays of certainty and/or uncertainty were not a preferred stylistic choice in none of the three datasets. The provision of details was likely considered more relevant by the reviewers especially when the explicit goal was that to persuade other travelers not to choose a certain hotel.

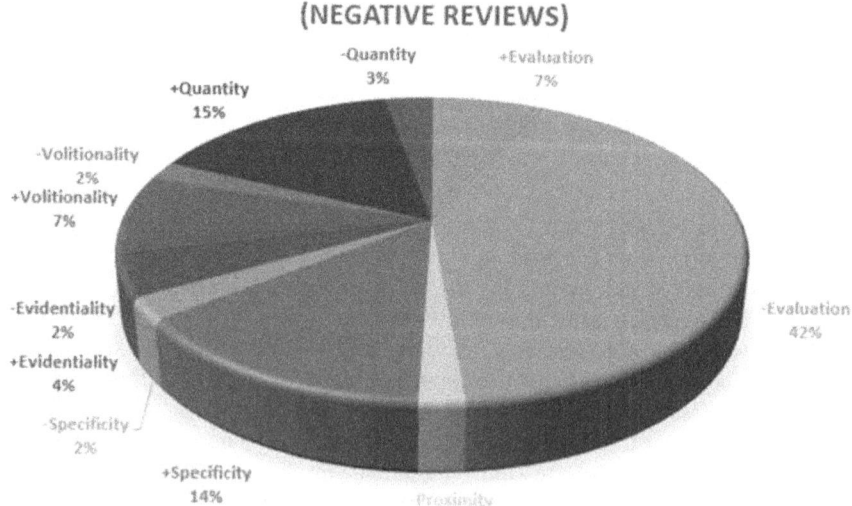

Figure 2. Distribution of the emotive devices in the negative reviews' dataset.

Hypoteses of analysis on the stylistics of arguments

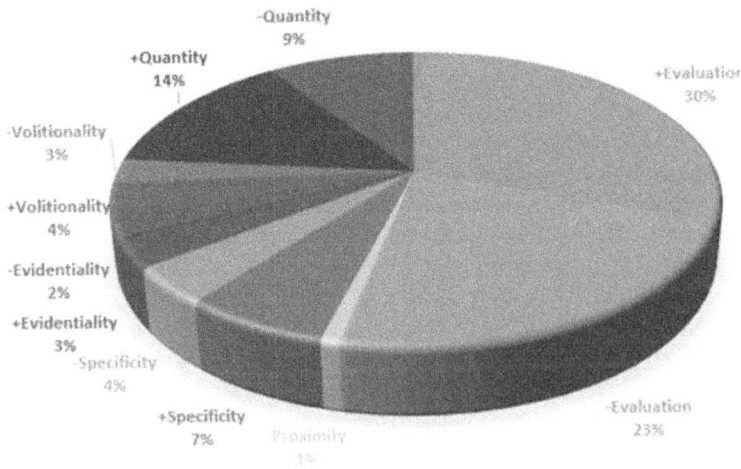

Fig. 3: distribution of the emotive devices in the average reviews' dataset.

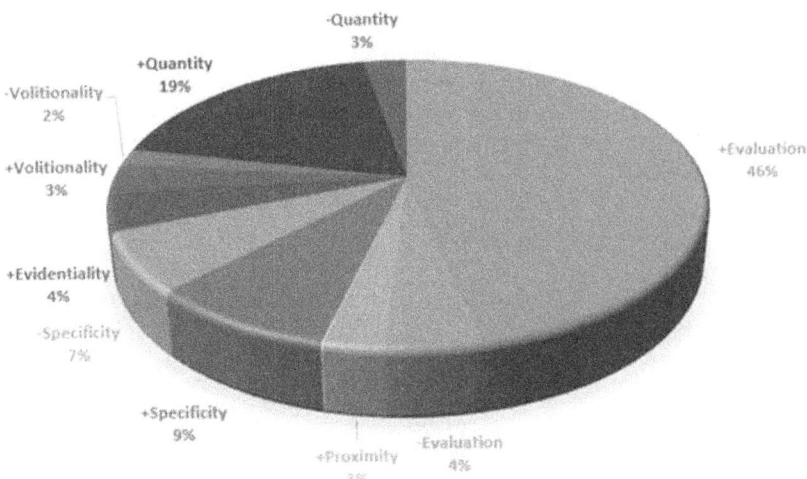

Figure 4. Distribution of the emotive devices in the positive reviews' dataset.

Unsurprisingly, evaluation devices were clearly the most preferred choice in all datasets. Linear correlations confirm effects of proximity ($r = 0,1895$; $r^2 = 0,0359$), vagueness ($r = 0,442$; $r^2 = 0,1954$), certainty ($r = 0,2158$; $r^2 = 0,0466$) and intensity ($r = 0,4421$; $r^2 = 0,1955$) in correlation with positive evaluations. Effects of distance ($r = 0,3315$; $r^2 = 0,1099$) and assertivity ($r = 0,3376$; $r^2 = 0,114$) were found in correlation with negative evaluations, as well as small effects of specificity ($r = 0,1383$; $r^2 = 0,0191$). Correlations of the specificity devices with the negative evaluations were unsurprisingly weaker than expected on their whole, considering that they might have been strategically employed differently in the medium and in the negative reviews. Even if in the former reviewers displayed generally unenthusiastic evaluations of the hotels' staff, locations and facilities, they were not necessarily interested in discouraging other travelers in choosing those hotels, thus avoiding to provide details in order to do so, unlike in the latter (see Figure 5). Moreover, in the case of negative reviews, most details in support of the negative evaluative standpoints were provided in formats of complaint story, which also tend to respond to self-expressive needs: if on the one hand, the opportunity to write a review on Trip Advisor offers a space where pro-social activities can be made concrete (*e.g.:* helping other travelers), on the other hand, the very same space offers an opportunity for self-disclosure, which serves personal goals. Especially in case of frustrating experiences, reviewers may also want to pour out and seek understanding on their ruined stays, other than giving advice to strangers. Diffused low correlations of negative assertivity (which most likely marks shields and other mitigating strategies) with positive and negative evaluations in all datasets, might be explained by the type of asynchronous and asymmetrical digital interaction that Trip Advisor provides to travelers: hotel managers, who have accessibility to their travelers reviews, may or may not respond to them; even if instances of flame are not rare, they still happen behind a screen, hence reducing face-saving needs in the case of negative evaluations. Generally non-mitigated positive evaluations may also lead to think of a more concrete and straight-to-the-point digital communicative style, which is used in a context which appears to be less exposing and less risky than face-to-face situations.

	Evaluation +				Evaluation -			
	r	r²	P (one-tailed)	P (two-tailed)	r	r²	P (one-tailed)	P (two-tailed)
Proximity+	0.1895	0.0359	0.0295285	0.059057	-0.2826	0.0799	0.0021715	0.004343
Proximity-	-0.3288	0.1081	0.02417	0.000828	0.3315	0.1099	0.000375	0.00075
Specificity+	-0.0114	0.0001	0.456317	0.912634	0.1383	0.0191	0.0853635	0.170727
Specificity-	0.442	0.1954	<.0001	<.0001	-0.0962	0.0093	0.169709	0.339418
Evidentiality+	0.2158	0.0466	0.0154485	0.030897	-0.155	0.024	0.0621825	0.124365
Evidentiality-	-0.1474	0.0217	0.0723825	0.144765	0.0788	0.0062	0.2186355	0.437271
Volitionality+	-0.1215	0.0148	0.1145945	0.229189	0.3376	0.114	0.000297	0.000594
Volitionality-	0.0225	0.0005	0.4131645	0.826329	0.002	0	0.492042	0.984084
Quantity+	0.4421	0.1955	<.0001	<.0001	-0.215	0.0462	0.015826	0.031652
Quantity-	0.2595	0.0673	0.004565	0.00913	0.0549	0.003	0.295211	0.590422

Figure 5. Emotive devices correlating with evaluation+/- in all three datasets (general means weighted on an average of 100 words).

7. Conclusions

Computer Mediated Communication can provide opportunities of goal-oriented argumentation by means of, *mutatis mutandis*, macro- and micro-stylistic strategies similar to those occurring in conversational settings. At a first glance, it even seems that manifestations of certain standpoints, especially those dealing with personal experiences and loaded with positive or negative affectivity, are more heavily communicated online than in face-to-face conversations (see also Bonelli 2015). The analyses I proposed are an attempt to detect different travelers' stylistic strategies used to present and reinforce their arguments while reviewing a group of hotels in London, UK. In particular, I tried to consider how evaluative standpoints could be detected and measured by means of markers such as Caffi & Janney (1994)'s emotive devices. The framework I used allowed me to consider how expressive and semantic levels of argumentation should not be seen as separated instances, insofar as the micro-stylistic choices speakers adopt (*e.g.*: the degrees of proximity, specificity, certainty, assertivity and intensity) can give us important information on how they try to make their standpoints more evident, as well as their commitment to them.

However, it is important to notice that this kind of perspective may vary depending on different topics, contexts, situations, registers and cultures. Finally, the proposed framework may best work as complementary of classic argumentative ones, rather than as a substitute of them. An interplay between argumentative frameworks that cast attention to the semantics-pragmatics interface (e.g.: Luciani 2014; Rigotti & Rocci 2006) and the one hereby proposed should posit itself in terms of focus selection of the units of analysis: rather than seeing the former approach as the only one dealing with core-meanings and the latter as the only one dealing with interpersonal strategies (*e.g.*: adopted style), seeing the former as the one dealing with macro-levels of analysis and the latter with micro-ones might be a potentially fruitful methodology to analyze online arguments, especially where a lower need of face-saving strategies (as in the case of asynchronous and asymmetric CMC as Trip Advisor reviews are) make commitments so explicit and clear-cut.

References

Abercrombie, D. (1967), *Elements of general phonetics*. Edinburgh: Edinburgh University Press.

Akhmanova, O. (1976), *Linguostylistics: Theory and Method*. The Hague: Mouton.

Angouri, J., Tseliga, T. (2010), *"You Have No Idea What You are Talking About!" From e-disagreement to e-impoliteness in two online fora* in «Journal of Politeness Research», 6: 57-82.

Arndt, H., Janney, R. W. (1987), InterGrammar. *Toward an Integrative Model of Verbal, Prosodic and Kinesic Choices* in Speech. Berlin, New York, Amsterdam: Mouton de Gruyter.

Bally, C. (1909), *Traité de stylistique française*. 2 Voll. Genève: Librairie de l'Université Georg & Cie S. A.

Benveniste, E. (1970), *L'appareil formel de l'énonciation* in «Langages», 17. Rist. in: Id., *Problèmes de linguistique générale, II*, 1974: 79-88. *Problemi di linguistica generale II*, Milano: Il Saggiatore, 1985: 96-106.

Blum-Kulka, S. (1987), *Indirectness and politeness in requests: Same or different?* in «Journal of Pragmatics», 11: 147- 160.

Bradac, J. J., Hosman, L. A., Tardy, C. H. (1978), *Reciprocal disclosures and language intensity: attributional consequences* in «Communication Monographs», 45: 1-17.

Briz, A. (1995), *La atenuación en la conversación coloquial. Una categoría pragmática* in Cortés, L.(ed.), 103-122.

Brown, P., Levinson, S. [1978] (1987), *Politeness. Some universals in language usage*. Cambridge: Cambridge University Press.

Bühler, K. [1934] (1999), *Sprachtheorie*. Stuttgart: Lucius & Lucius.

Caffi, C. (2007), *On Mitigation*. Bingley: Emerald.

—— (2006), *Metapragmatics* in Brown, K. (editor-in-chief) Language & Linguistics, Second Edition, Vol. 8, 82-88. Oxford: Elsevier.

—— (2001), *La Mitigazione. Un approccio pragmatico alla comunicazione nei contesti terapeutici*. Münster, Hamburg, London: LIT Verlag.

Caffi, C., Janney, R. W. (1994), *Toward a pragmatics of emotive communication* in «Journal of Pragmatics», 22: 325-373.

Chmiel, A., Sienkiewicz, J., Thelwall, M., Paltoglou, G., Buckley, K. (2011), *Collective Emotions Online and Their Influence on Community Life* in «PLoS One», 6(7): e22207.

Clark, H. H. (1996), *Using Language*. Cambridge: Cambridge University Press.

Conte, M. E. (1990), *Anaphore, Predication, Empathie* in Charolles, M., Fisher, S., Jayez, J. (eds.), 215-225.

Delbouille, P. (1960), *A propos de la définition de du fait de style* in Cahiers d'Analyse Textuelle, 2: 94-104.

Du Bois, J. W. (2007), *The stance triangle* in Englebretson, R. (ed.), 140-182.

Englebretson, R. (ed.) (2007), *Stancetaking in discourse: subjectivity, evaluation, interaction*. Amsterdam: John Benjamins.

Fabri, M., Moore, D. J., Hobbs, D. J. (2005), *Empathy and enjoyment in instant messaging* in Proceedings of 19th British HCI Group Annual Conference (HCI2005), Edinburgh, UK, 4-9.

Fiorentino, G. (in print), *Forme di scrittura in rete: dal web 1.0 al web 2.0*.

Fraser, B. (1980), *Conversational Mitigation* in «Journal of Pragmatics», 4: 341-350.

Frijda, N. H. (1982), *The meanings of emotional expressions* in Key, M. R. (ed.), 103-119.

Giles, H., Coupland, J., Coupland, N. (eds.) (1991), *Contexts of accommodation: Developments in applied sociolinguistics*. Cambridge: Cambridge University Press.

Giles, H., Scherer, K. R., Talor, D. M. (1979), *Speech markers in social interaction* in Scherer, K., Giles, H. (eds.), 343-381.

Gill, A. J., Gergle, D., French, R.M., Oberlander, J. (2008), *Emotion Rating from Short Blog Texts*. Proceedings of the ACM Conference on Human Factors in Computing Systems (CHI 2008): 1121-1124.

Goodwin, C. (2007), *Participation, stance, and affect in the organization of activities* in «Discourse and Society», 18: 53-73.

Goodwin, M. H., Cekaite, A., Goodwin, C. (2012), *Emotion as Stance* in Peräkylä, A., Sorjonen, M. L. (eds.), 16-41.

Goffman, E. (1971), *The territories of the self.* in «Relations in public», 28-61.

Gough, H. H. (ed.) (1957), *Manual for the California psychological inventory*. Palo Alto, CA: Consulting Psychologist Press.

Gretzel U., Yoo, K. H. (2008), *Use and impact of online travel reviews*. In P. O'Connor, W. Höpken & U. Gretzel (Eds.), *Information and Communication Technologies* in Tourism 2008 (pp. 35-46). Vienna, Austria: Springer-Verlag Wien.

Grice, H. P. (1989), *Studies in the Way of Words*. Cambridge: Harvard University Press.

(1975), *Logic and Conversation* in Cole, P., Morgan, J. L. (eds.), 41-58.

Gumperz. J. J. (1982), *Discourse Strategies*. Cambridge: Cambridge University Press.

Hancock, J. T., Landrigan, C., Silver, C. (2007), *Expressing emotion in text-based communication*. Proc ACM Conference on Human Factors in Computing Systems (CHI 2007): 929-932.

Havertake, H. (1992), *Deictic categories as mitigating devices* in «Pragmatics», 2: 505-522.

Hickey, L. (ed.) (1989), *The Pragmatics of Style*. London: Routledge.

Holmes, J. (1984), *Modifying illocutionary force* in «Journal of Pramatics», 8: 345-365.

Hölker, K. (1988), *Zur Analyse von Markern*. Stuttgart: Franz Steiner.

Hübler, A. (1987), *Communication and expressivity* in Dirven, R., Fried, V. (eds.), 357-380.

Jaffe, A. M. (ed.) (2009), *Stance: sociolinguistic perspectives*. Oxford: Oxford University Press.

Janney, R. W. (1996), *E-mail and Intimacy* in Gorayska, B., Mey, J. L. (eds.), *Cognitive Technology: in Search of a Human Interface*. Amsterdam: Elsevier Science B. V., 201-208.

Kärkkäinen, E. (2006), *Stance taking in conversation: from subjectivity to intersubjectivity* in «Text & Talk», 26: 699-731.

(ed.) (2003), *Epistemic Stance in English Conversation: A Description of its Interactional Functions, with a Focus on 'I think'*. Amsterdam: John Benjamins.

Kerbrat-Orecchioni, C. (ed.) (1992), *Les Interactions verbales, tome II*. Paris: A. Colin.
Kleinke, S. (2008), *Emotional Commitment in Public Political Internet Message Boards* in «Journal of Language and Social Psychology», 27: 409-421.
Labov, W. (1984), *Intensity* in Schiffrin, D. (ed.), 43-70.
Labov, W., Fanshel, D. (1977), *Therapeutic discourse: psychotherapy as conversation*. New York: Academic Press.
Lakoff, R. (1974), *Remarks on this and that* in Papers from the Tenth Regional Meeting of the Chicago Linguistic Society, 345-356. Chicago: University of Chicago Press.
Langlotz, A., Locher, M. A. (2012), *Ways of communicating emotional stance in online disagreements* in «Journal of Pragmatics», 44: 1591-1606.
Leech, G. (1983), *Principles of pragmatics*. London: Longman.
   (1980), *Explorations in Semantics and Pragmatics*. Amsterdam, Philadelphia: John Benjamins Publishing Company.
Locher, M., Graham, S. L. (2010), *Handbook of Pragmatics, Vol. Interpersonal Pragmatics*. Berlin, New York: De Gruyter Mouton.
Luciani, M. (2014), *The evaluative and unifying function of emotions emerging in argumentation: interactional and inferential analysis in highly specialized medical consultations concerning the disclosure of a bad news*. Proceedings of the ISSA Conference, Amsterdam.
Lyons, J. [1981] (1986), *Language, Meaning and Context*. Bungay: Fontana Paperbacks.
   (1977), *Semantics. Vol 2*. Cambridge: Cambridge University Press.
Merlini Barbaresi, L. (2009), *Linguaggio intemperante e linguaggio temperato. Ovvero intensificazione arrogante e attenuazione cortese* in Gili Fivela, B., Bazzanella, C. (eds), 59-78.
Meyer-Hermann, R., Weingarten, R. (1982), *Funktion einer Äußerung. Sprache erkennen und verstehen* in Akten des 16 Linguistischen Kolloquiums, Kiel 1981, 2, 242.
Marwick, A. E., Boyd, D. (2010), *I tweet honestly, I tweet passionately: Twitter users, context collapse, and the imagined audience* in «New media & Society», XX(X): 1-20.
Miceli, M., Castelfranchi, C. (2014), *Expectancy and emotion*. Oxford: Oxford University Press.
Mortara Garavelli, B. (1988), *Manuale di retorica*. Milano: Bompiani.
Niemelä, M. (2010), *The reporting space in conversational storytelling: Orchestrating all semiotic channels for taking a stance* in «Journal of Pragmatics», 42: 3258-3270.
Ochs, E., Capps, L. (1996), *Narrating the Self* in «Annual Review of Anthropology», 25: 19-43.
Ochs, E., Schieffelin, B. B. (1989), *Language has a heart* in Ochs, E. (ed.), *The pragmatics of affect* in «Text», 9: 7-25.
Orletti, F. (ed.) (2004), *Scrittura e nuovi media. Dalle conversazioni in rete alla Web usability*. Roma: Carocci.

Osgood, C. E., Suci, G. J., Tannenbaum, P. H. (1957), *The Measurement of Meaning*. Urbana, Chicago, London: University of Illinois Press.

Perelman, C., Olbrechts-Tyteca, L. [1958] (1966), *Traité de l'argumentation. La nouvelle rhétorique*. Paris: Presses Universitairies de France. Trad. it. *Trattato dell'argomentazione*. Torino: Einaudi.

Pistolesi, E. (ed.) (2002), *Flame e coinvolgimento in IRC (Internet Relay Chat)* in Bazzanella, C., Kobau, P. (eds), 261-277.

Provine, R. R., Spencer, R. J., Mandell, D. L. (2007), *Emotional expression online emoticons punctuate website text messages* in «Journal of Language and Social Psychology», 26: 299-307.

Riffaterre, M., (1960), *Stylistic Context* in «Word», 16: 39-55.

Rigotti, E., Greco Morasso, S. (2010), *Comparing the Argumentum Model of Topics to other contemporary approaches to argument schemes: the procedural and material components* in «Argumentation», 24(4), 489-512.

Rigotti, E., Rocci, A. (2006), *Tema-rema e connettivo: la congruità semantico-pragmatica del testo* in Gobber, G., Gatti M. C., Cigada, S. (Eds.) Syndesmoi (pp. 1000-1044), Milan: Vita e Pensiero.

Schiffrin, D. (1987), *Discourse markers*. Cambridge: Cambridge University Press.

Schneider, S. (2010), *Mitigation* in Locher, M. L., Graham, S. L. (eds.), 253-269.

Selting, M. (2010), *Affectivity in Conversational Storytelling* in «Pragmatics», 20: 229-277.

(2000), *The construction of units in conversational talk* in «Language in Society», 29: 477-517.

Tannen, D. [1989] (2007), *Talking Voices. Repetition, Dialogue, and Imagery* in Conversational Discourse. Cambridge: Cambridge University Press.

Thelwall, M., Buckley, K., Paltoglou, G. (2011), *Sentiment in Twitter events* in «Journal of the American Society for Information Science and Technology», 62: 406-418.

Thelwall, M., Wilkinson, D., Uppal, S. (2010), *Data mining emotion in social network communication: Gender differences in MySpace* in «Journal of the American Society for Information Science and Technology», 21: 190-199.

Venier, F. (2008), *Il potere del discorso. Retorica e pragmatica linguistica*. Roma: Carocci.

Verschueren, J. (2000), *Notes on the role of metapragmatic awareness in language use* in «Pragmatics», 10: 439-456.

Volek, B. (1987), *Emotive signs in language and semantic functioning of derived nouns in Russian*. Amsterdam, Philadelphia: John Benjamins Publishing Company.

Yus, F. (2013), *Cyberpragmatics. Internet-mediated communication in context*. Amsterdam: John Benjamins.

Part III
Tools and Applications

Chapter 9

ProtOCL: Specifying Dialogue Games Using UML and OCL

Tangming Yuan[1] and Simon Wells[2]

[1] University of York, York, United Kingdom, tommy.yuan@york.ac.uk
[2] Edinburgh Napier University, Edinburgh, United Kingdom, s.wells@napier.ac.uk

Abstract. Dialogue games are becoming increasingly popular tools for Human-Computer Dialogue and Agent Communication. However, whilst there is an increasing body of theoretical underpinning that demonstrates the value and utility of dialogue games, and also a range of novel implementations within specific problem domains, there remain very few tools to support the deployment of dialogue games based solutions within new problem domains. This paper introduces a new approach, called ProtOCL, to the specification of dialogue games. This approach adopts Unified Modeling Language (UML) and the Object Constraint Language (OCL) and enables the rapid movement from specification to deployment and execution. The dialogue game, DE, is used as an exemplar and is described using OCL to yield DE-OCL. Code generation is subsequently used to move from the DE-OCL description to executable code. This approach goes beyond existing description languages and their supporting tools by (1) using a description language that is familiar to a far larger user group, and, (2) enabling code-generation using languages and technologies that are current industry standards.

1. Introduction

Dialectics is a branch of philosophy that seeks to build models for "fair and reasonable" dialogue (Walton & Krabbe, 1995). A common approach within dialectics is to construct dialogue games such as those of Hamblin (Hamblin, 1970), Walton and Krabbe (Walton & Krabbe, 1995), Walton (Walton, 1998), and Mackenzie (Mackenzie, 1990). A dialogue game can be seen as a prescriptive set of rules, regulating the participants as they make moves in a dialogue. These rules legislate as to the permissible sequences of moves, and also as to the effect of moves on the participants' "commitment stores", a record of the player's positions with respect to the statements made thus far. Such dialogue games have received much recent interest from people working in Human Computer Dialogue and in Artificial Intelligence, for example Bench-Capon and Dunne (Bench-Capon & Dunne, 2007), Reed and Grasso (Reed & Grasso, 2007), Rahwan and McBurney (Rahwan & McBurney, 2007), and Yuan et al.(Yuan, Moore, Reed, Ravenscroft, & Maudet, 2011).

A number of computerised dialectical systems have been designed. Grasso et al. (Grasso, Cawsey, & Jones, 2000) for example, outline a system designed to change the attitudes of its users in the domain of health promotion. Vreeswijk (Vreeswijk, 1995) has designed "IACAS", an interactive argumentation system enabling disputes between a user and the computer. Yuan et al. (Yuan, Svansson, Moore, & Grierson, 2007) have applied the argument game from Wooldridge (Wooldridge, 2002) and the abstract argumentation system of Dung (Dung, 1995) to the construction of a computational argument game called "Argumento". Argumento enables human-agent, agent-agent and human-agent to ex- change both abstract and concrete arguments (Yuan & Schulze, 2008) and has been adopted as the core for an arguing agents competition (Yuan, Schulze, Devereux, & Reed, 2008; Wells, Lozinski, & Pham, 2008). Ravenscroft and Pilkington (Ravenscroft & Pilkington, 2000) used a dialogue game framework to facilitate a "structured and constrained dialectic" which in turn aided the student in enhancing explanatory domain models in ways that led to conceptual development concerning the physics of motion. The framework has been implemented in a prototype system "CoLLeGE" (Computer based Lab for Language Games in Education). Empirical studies have shown the effectiveness of the dialogue game framework (Ravenscroft, 2000; Ravenscroft & Matheson, 2002). Mackenzies dialectical system named 'DC' (Mackenzie, 1979) has been used as the basis for developing a further system named 'DE' (Yuan, Moore, & Grierson, 2003) which has been used as the underlying model for a human-computer debating system (Yuan, 2004; Yuan, Moore, & Grierson, 2007, 2008). Recently, dialogue games have also been used to structure interaction between humans and intelligent agents in mixed initiative environment as demonstrated in the MultiAgent Argument Logic and Opinion (MAgtALO) systems (Reed & Wells, 2007) and between intelligent agents within a multi-agent system (Kalofonos et al., 2006). The systems we have outlined above face a distinct formal representation problem that is the representation of the structure of the protocol that governs the dialogue game as it unfolds (Yuan et al., 2011). We may, for example, build a system that is to use the DE model to argue about capital punishment, but how are we to store the rules of DE?

To date, computational dialectic systems have approached the problems by expressing dialogue models informally using plain or structured English and then hard-wiring them into the program structure by the developer of the system. Hard-wiring means that the game rules cannot be easily modified unless re-coded and the entire system rebuilt. This makes reuse impossible as the game rules cannot be formally specified, saved and subsequently interfaced by other systems. A formal means of specifying dialogue games is therefore needed. In this paper we report on an approach to the specification of dialogue games that we have named ProtOCL. This approaches uses the Unified Modeling Language (UML) to describe a generic dialogue game consisting of the common core elements found across a range of dialogue games, and the Object Constraint Language (OCL) to express specific rules as UML annotations that enable the generic game to be made specific to a particular dialogue game.

The remainder of this paper is organised as follows. Section 2 provides literature reviews of different methods for specifying agent dialogue protocols. Section 3 argues and demonstrates the case of using of UML and OCL as an approach to specify dialogue games. Section 4 discusses how a dialogue game framework can be generated and interfaced by the dialogue game engine and agents. Section 5 concludes the paper and point out our intended future work.

2. Methods for Specifying Dialogue Protocols

This section reviews some of the methods from the literature that have been used to specify dialogue protocols. These generally fall into the following categories: (i) natural language, (ii) formal logical notation, (iii) diagrammatic, and, (iv) domain specific language (DSL).

Natural languages descriptions, such as the game DE, a simplified version of which is illustrated as follows:

Move Types

Assertions: The content of an assertion is a statement P, Q, etc. or the truth-functional compounds of statements: "Not P", "If P then Q", "P and Q".

Questions: The question of the statement P is "Is it the case that P?"

Challenges: The challenge of the statement P is "Why P?"

Withdrawals: The withdrawal of the statement P is "no commitment P".

Resolution demands: The resolution demand of the statement P is "resolve whether P".

Dialogue Rules

R_{FORM} : Participants may make one of the permitted types of move in turn.

$R_{REPSTAT}$: Mutual commitment can only be asserted when a question or challenge is responded.

R_{QUEST}: The question P can be answered only by P, "Not P" or "no commitment P".

R_{CHALL}: "Why P?" has to be responded to by either a withdrawal of P, a statement that the challenger accepts, or a resolution demands of the previous commitments of the challenger which immediately imply P.

$R_{RESOLVE}$: A resolution demand can be made only in situations that the other party of the dialogue has committed in an immediate inconsistent conjunction of statements, or he withdraws or challenges an immediate consequent of previous commitments.

$R_{RESOLUTION}$: A resolution demand has to be responded by either the withdrawal of the offending
conjuncts or confirmation of the disputed consequent.
$R_{LEGALCHALL}$: "Why P?" cannot be used unless P has been explicitly stated by the dialogue partner.

Commitment Rules

Initial commitment, CR_0: The initial commitment of each participant is null.
Withdrawals, CR_W: After the withdrawal of P, the statement P is not included in the move makers store.
Statements, CR_S: After a statement P, unless the preceding event was a challenge, P is included in the move makers store.
Defence, CR_{YS}: After a statement P, if the preceding event was Why Q?, P and If P then Q are included in the move makers store.
Challenges, CR_Y: A challenge of P results in P being removed from the store of the move maker if it is there.

Termination Rules

1. The game will be ended when a participant accepts another participants view.

Such descriptions are both popular and plentiful in the literature, are generally well organized but have drawbacks, most importantly from the computation perspective, failing to lend themselves to either immediate execution or automated evaluation. In a natural language description, the rules of the game are generally grouped into a limited number of categories that define the types of available move (locution rules), how the moves interact with each other (structural rules), how playing the moves affect the commitments of the players (commitment rules), and the circumstances under which the game comes to an end (termination rules). A strength of the natural language approach is that the resulting descriptions are expressive and are, to a degree, easily understood by developers. However natural language descriptions of game rules can lead to problems with ambiguity. This aspect is compounded when the aim is for computational use of the game as natural language specifications are generally not machine-readable so automated testing, deployment, and execution become a difficult problem.

Formal specifications use notations from mathematical or formal logics to represent the semantics of dialogue rules in a precise way. Notable examples of this approach are to be found in (Bodenstaff, Prakken, & Vreeswijk, 2006), (Prakken, 2005) (Brewka, 2001) and (Artikis, Sergot, & Pitt, 2007). Recent work has attempted to create more generic description formats that retain the rigorousness of the formal approach whilst providing a range of descriptive features that are closer to the problem domain, for example in (Wells & Reed, 2004), the typical moves of Hamblin-type dialectical games are characterised in terms of a limited number of states and updates. The rules of a complete game are then expressed by assembling collections of moves in terms of pre- and post- conditions using a set-theoretic formal notation. For example the following set theoretic specification for the Hamblin-type game illustrates some of the pre-condition checks on commitment store content:

$C \in CSn$ Commitment C is currently in commitment Store CS

$C \notin CSn$ Commitment C is not currently in commitment Store CS

The following specification illustrates some of the post-conditions for commitment store alterations:

$CS_{n+1} = CS_n \cup \{C\}$ Commitment C is added to Commitment Store CS
$CS_{n+1} = CS_n \cap \{C\}$ Commitment C is removed from Commitment Store CS

The following example then uses the expressions from for the commitment store checks and updates to complete the pre- and post-conditions for the statement and withdrawal moves of Hamblin's game, H:

Statement(S_x)   Pre: $\emptyset$
                 Post: $CP_{n+1} = CP_n \cup \{S_x\} \wedge CO_{n+1} = CO_n \cup \{S_x\}$

Withdrawal(S_x)   Pre: $\emptyset$
                 Post: $CP_{n+1} = CP_n \setminus \{S_x\}$

Whilst this approach improves the specificity of the rules and leads to a reduced chance for ambiguity, this approach is difficult to communicate to developers who do not possess the necessary mathematical background required to understand the notation (cf. Sommerville, 2011).

A Domain specific language (DSL) provides an intermediate position between the natural-language and formal approaches. An aim of the DSL approach is to re-use the established language of the domain problem, e.g. language used by people working with dialectical games, so that developers have an intuitive understanding of the expressions in the language, but to confine the expressions to those that are legal according to a formal grammar. Thus protocols are expressed in a way that is executable, assuming that adequate tooling is created to support the language, and immediately comprehensible to those versed in the language of the problem domain. An example of this kind of approach can be found in the Dialogue Game Description Language (DGDL) (Wells & Reed, 2012) a DSL that is founded on an Extended Backus-Naur Form (EBNF) grammar[1] to support the description of

---

[1] https://github.com/siwells/DGDL/tree/master/grammar

syntactically correct and verifiable dialectical games. The language at the current stage of development, however, needs software tool support particularly in terms of user-facing (design) tools and execution "engines". The following is an example of a DGDL game description name "Simple":

```
Simple{
    {turns,magnitude:single,ordering:strict}
    {players,min:2,max:2}
    {player,id:Player1}
    {player,id:Player2}
    {store,id:CStore,owner:Player1}
    {store,id:CStore,owner:Player2}
    {Assert,{p},"I assert that",
        {store(add, {p}, CStore, Speaker),store(add, {p}, CStore, Listener)}
    }
}
```

In this example game a turn structure, two named players, and a commitment store for each player are defined. A single assert move is then defined which incurs commitment in both players commitment stores when it is played. This game is for purely illustrative purposes and is indicative of the features and descriptive character of DGDL descriptions.

There have been a variety of approaches to the diagrammatic description of dialogue protocols. For example, in the Toulmin Dialogue Game (TDG) (Bench-Capon, 1998) a state diagram is used to regulate the order of moves and assignment of roles within a TDG dialogue. Finite State Machines (FSMs) have long been used to define network protocols and have been widely used to model, analyse, and prototype distributed systems (Shen, Norrie, & Barthes, 2001). FSMs have also been used to describe conversation policies in multi-agent systems (Bradshaw, Dutfield, Benoit, & Woolley, 1997). UML sequence diagrams also provide a way to diagrammatically depict dialogue protocols. For example, Agent UML (AUML) (Odell, Bauer, & Parunak, 2001) extends the unified modeling language (UML) to model intelligent software agents and related agent-based systems. FIPA adopted this approach to specify agent communication protocols such as the Subscribe Interaction Protocol[2], which enables an agent to subscribe to messages from another agent with respect to a specific referenced object. The state machine and sequence diagram approach may be more suitable for simple communication protocols (Norman, Carbogim, Krabbe, & Walton, 2004) as they visualize the actually occurred sequence of communications. It is not clear how certain constraints and rules, for example the DE rule $R_{REPSTAT}$: Mutual commitment may not be asserted unless to answer a question or a challenge can be represented on the diagram.

In summary, natural language specifications are not machine readable and are subject to ambiguity, formal approaches are generally not user- and developer-

---

[2] http://www.fipa.org/specs/fipa00035/

friendly, diagrammatic approaches are suitable for more simple protocols but must be underpinned by some formal representation to enable them to be executable without additional work, and DSLs currently have insufficient tooling to support wide-scale popular adoption.

3. Specifying Dialogue Games Using UML

The Unified Modeling Language (UML) is a language for specifying, visualizing, constructing, and documenting the artifacts of software systems. The latest version, UML 2.0, supports 13 types of diagrams including class, sequence, and state diagrams and one language; the object constraint language (OCL). OCL is used to describe rules that apply to UML models and adds vital information to the model that cannot be otherwise depicted using diagrammatic means. OCL is formally defined, readable and writeable by both humans and a range of software tools, and is easy to use (O.M.G., 2012). This goes a long way towards satisfying the need for a developer friendly formal language for describing dialogue games.

To use UML to represent the rules of a dialogue game, a model that captures general properties of a dialogue game, such as that depicted in the UML class diagram shown in Figure 1, is constructed. The class diagram captures common terms in the dialogue game domain such as the: game, player, dialogue history, commitment store, turn, move, move content, proposition and inference. Each dialogue game has a thesis and two players, a proponent of the thesis and an opponent. Each game also contains a dialogue history that records the moves made by the players on a turn-by-turn basis. The size of the dialogue history is the total number of turns made by both players. Each turn has a unique number and a player may make one or more moves in one turn. Each move contains a move type and a move content, which could be a proposition or an inference. Each inference contains a set of data and a conclusion. The negative value of a proposition or inference can be retrieved on request. Each player has a commitment store that record the statements made or accepted during dialogue. The commitment stores are publicly inspectable so a player can also view their opponent's store. The internal structure of a commitment store can be flexible depending on individual games, e.g. to maintain separate lists of propositions or inferences. A proposition or an inference can be checked against a commitment store to see whether it is supported by others or by itself. The latter is useful for banning circular arguments in dialectical systems. While the class diagram specifies the generic terms used by dialogue games, OCL is required to annotate the class diagram in order to provide a full specification of a dialogue game. The description of DE as presented in 1 is used to demonstrate this. For example, the DE move types rule can be specified as

```
--Player makes a legal move
context Player::makeMove():Move
    --Permitted move types:
```

> post: Set{'Assertion', 'Question', 'Challenge', 'Resolve', 'Withdrawal'}
> ->includes(result.getType())

The rule is specified as a post condition within the context of player make-Move operation. context, post and result are OCL keywords and includes is an OCL operation that applies to a set.

The DE dialogue rule RFORM can be specified as

> --RFORM: Participants may make one of the permitted types of move in turn.
> context Turn
>   inv: move->size()=1
> context DialogueGame
>   inv: self.proponent.turn->forAll(getNumber()/2=1) and self.opponent.turn
>   ->forAll(getNumber()/2=0)

The rule is specified jointly within the context of Turn and DialogueGame class as two invariants: the first is that the set of moves associated with each turn is exactly one and the second is that the turn numbers for the proponent are odd numbers and for the opponent are even numbers given that the proponent always starts a game. 'inv', 'self', and 'and' are all OCL keywords and size is an OCL operation that applies to a set.

The DE commitment rule CR0 can be specified as

> --Initial commitment, CR0: The initial commitment of each participant is null.
> context DialogueGame::start():String
>   post: proponent.store.content->isEmpty() and
>   opponent.store.content->isEmpty()

The rule is specified within the context of dialogue game start operation as post conditions. isEmpty is an OCL operation that applies to a set.

The DE termination rule can be specified as

> --Termination Rules: The game will be ended when a participant accepts the other participant's view.
> context DialogueGame::end():String
>   pre:proponent.store.content->includes(thesis.getNegation()) or
>   opponent.store.content->includes(thesis)
> --Playing
> context DialogueGame::play():String
>   pre: proponent.store.content->excludes(thesis.getNegation())and
>   opponent.store.content->excludes(thesis)

The precondition for a dialogue game to end is that one party's store contains the opponent's thesis. Otherwise, the game is in the playing state.

4. Automatic Generation of Dialogue Game Framework

Given a description of a dialogue game, such as the description of DE as presented in Section 1, and a UML diagram of a generic dialogue game as previously depicted in Figure 1, the UML diagram can be annotated using an OCL specification file. An example of such a file can be found in the protocol_de_ocl description file[3], which presents the OCL expressions used to describe DE. Suitable tools can subsequently process this description file. The Object Constraint Language Environment (OCLE)[4] is one example, to generate executable code. The output of the code generation stage is executable Java code. This process essentially yields the core of a dialogue game engine via code generation, for example, providing checks against the dialogue rules when a particular operation, such as makeMove, is invoked, and generating an error message if a player breaks the game rules.

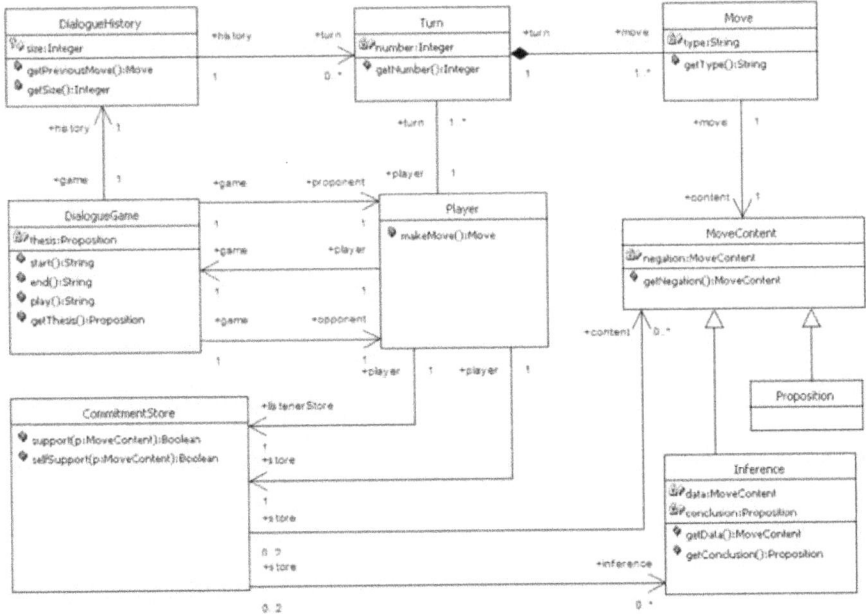

Figure 1. A generic model for dialogue games expressed using the UML class diagram notation. This model captures the general core elements of dialogue games and provides the basis for an API for generated code.

---

[3]This is available from the ProtOCL Git repository located at the following URL: https://github.com/siwells/ProtOCL
[4] http://lci.cs.ubbcluj.ro/ocle/

There are a number of advantages to using code generation to pro- duce the core of the game engine. Primarily, it reduces the opportunity for the game designer to introduce errors. Additionally code-generation reduces the time required to implement and deploy new game engines and streamlines the effort required, by automating much of the implementation process. As a result effort can be focused on the design of the game rather than implementation details.

Because the game engine Application Programming Interface (API) is based upon the classes generated from the UML class diagram, which is static, it is straightforward to build new, dialogue aware tools against it, for example, using the engine to provide dialogue game managements for intelligent agents. This relationship is illustrated in Figure 2. Additionally, so long as the class model API remains unchanged, different games can be generated by OCL tools and then played by the agents via the game engine.

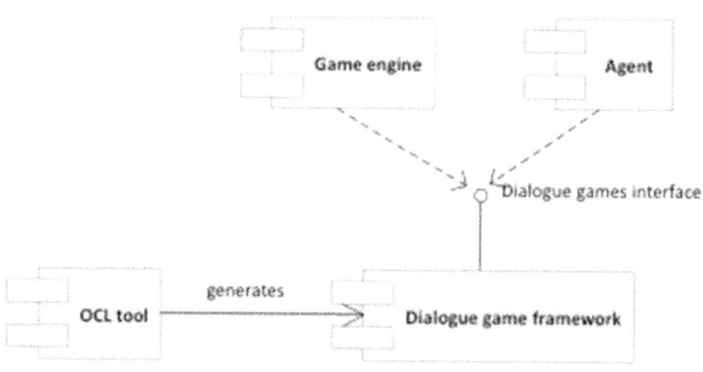

Figure 2. An overview of the ProtOCL dialogue system API from the code generation perspective. The OCL tool generates a dialogue game framework that exposes a common interface. This interface is then exploited by software within specific problem domains.

5. Conclusions & Future Work

Formal specification of dialogue games in a developer-friendly manner is attractive in terms of developing and testing dialogue games. OCL as part of the UML is a formal language to describe complex business rules and thus providing a tool framework that is more familiar to existing developers. The system we have introduced, ProtOCL, demonstrates how to use OCL to specify dialogue games via a generic UML class model that contains the terms and languages that are persistent to the dialectical system domain. Particularly, the naturally expressed dialogue games rules can be translated into OCL invariants and pre- and post- conditions. The expressive power of UML/OCL to the dialogue games has so far been demonstrated by the representation of the DE game using UML/OCL. A dialogue game framework can be generated via the existing OCL tools to generate the dialogue framework. The game engine and intelligent agents can then interface with this framework. Using this approach it would be convenient for the dialogue game designers to modify the game rules they are designing and test their games via the game engine and agents on the fly. It is anticipated that the proposed work represents a step forward in the implementation of dialogue games and dissemination of this approach within the wider software engineering context.

We plan to experiment with the generic class model enhancing it to support a wider variety of dialogue games to enable further refinements taking place. A dialectical system test-bed (the game engine and agents) can then be constructed for the game developers to test the dialogue games and dialogue strategies they have developed. One area of future work is to connect the current approach, using UML and OCL specify a dialogue game using human-oriented, graphical tools, with previous work on the Dialogue Game Description Language (DGDL) which enables games to be formally defined and syntactically verified. This opens the door to bi-directional movement of dialogue game specifications between a system that aims for increased human utility, and a system that aims for verifiability and formal correctness.

By bringing both approaches together we believe that a usable dialogue game definition and execution system can be assembled that enables software developers to build systems using the tools that they are already familiar with, and whose outputs can be evaluated and checked to ensure that they conform with developer expectations. This work contributes not just to the design and implementation of new games but also enables developers to immediately utilise dialogue games using industry standard software engineering tools and techniques.

References

Artikis, A., Sergot, M. J., & Pitt, J. (2007). An executable specification of a formal argumentation protocol. *Artificial Intelligence*, 171(10–15), 776–804.

Bench-Capon, T. J. M. (1998). Specification and implementation of Toulmin dialogue game. In *Proceedings of jurix 98* (p. 5-20).

Bench-Capon, T. J. M., & Dunne, P. E. (2007). Argumentation in artificial intelligence. *Artificial Intelligence*, 171(10–15), 619–641.

Bodenstaff, L., Prakken, H., & Vreeswijk, G. (2006). On formalising dialogue systems for argumentation in the event calculus. In Proceedings of the eleventh international workshop on nonmonotonic reasoning (pp. 374–382).

Bradshaw, J. M., Dutfield, S., Benoit, P., & Woolley, J. D. (1997). KAoS: Toward an industrial-strength generic agent architectiure. In *Software agents*. In (pp. 375–418). MIT Press, Cambridge, MA, USA.

Brewka, G. (2001). Dynamic argument systems: A formal model of argumentation processes based on situation calculus. *Logic and Computation*, 11(257–282), 619–641.

Dung, P. M. (1995). On the acceptability of arguments and its fundamental role in nonmonotonic reasoning and logic programming and n-person games. *Artificial Intelligence*, 77, 321-357.

Grasso, F., Cawsey, A., & Jones, R. (2000). Dialectical argumentation to solve conflicts in advice giving: a case study in the promotion of healthy nutrition. *International Journal of Human-Computer Studies*, 53 (6), 10771115.

Hamblin, C. L. (1970). *Fallacies*. Methuen and Co. Ltd.

Kalofonos, D., Karunatillake, N., Jennings, N. R., Norman, T. J., Reed, C., & Wells, S. (2006). Building agents that plan and argue in a social context. In P. E. Dunne & T. J. M. Bench-Capon (Eds.), *Computational models of argument* (pp. 15–26). IOS Press.

Mackenzie, J. D. (1979). Question begging in non-cumulative systems. *Journal Of Philosophical Logic*, 8, 117-133.

Mackenzie, J. D. (1990). Four dialogue systems. *Studia Logica*, 49, 567-583.

Norman, T. J., Carbogim, D. V., Krabbe, E. C. W., & Walton, D. (2004). Argument and Multiagent Systems. In *Argumentation machines: New frontiers in argument and computation*. Kluwer.

Odell, J., Bauer, B., & Parunak, H. V. D. (2001). Representing agent interaction protocols in UML. In 22nd international conference on software engineering (isce), berlin (p. 121-140).

O. M. G. (2012). Object constraint language (OCL) 2.3.1 specification, ISO/IEC 19507.

Prakken, H. (2005). Coherence and flexibility in dialogue games for argumentation. *Journal of Logic and Computation*, 15, 1009–1040.

Rahwan, I., & McBurney, P. (2007). Argumentation technology: introduction to the special issue. *IEEE Intelligent Systems*, 22(6), 21–23.

Ravenscroft, A. (2000). Designing argumentation for conceptual development. *Computers and Education*, 34, 241–255.

Ravenscroft, A., & Matheson, P. (2002). Developing and evaluating dialogue games for collaborative e-learning. *Journal of Computer Assisted Learning*, 18, 93–101.

Ravenscroft, A., & Pilkington, R. M. (2000). Investigation by design: developing dialogue models to support reasoning and conceptual change. *International Journal of Artificial Intelligence in Education*, 11, 273–298.

Reed, C., & Grasso, F. (2007). Recent advances in computational models of argument. International Journal of Intelligent Systems, 22(1), 1–15. Reed, C., & Wells, S. (2007). Using dialogical argument as an interface to complex debates. IEEE Intelligent Systems Journal: Special Issue on Argumentation Technology, 22(6), 60–65.

Shen, W., Norrie, D. H., & Barthes, J. (2001). *Multi-agent systems for concurrent intelligent design and manufacturing*. CRC Press.

Vreeswijk, G. A. W. (1995). IACAS: An implementation of Chisholm's principles of knowledge. In *proceedings of the 2nd Dutch/German workshop on non-monotonic reasoning*, Utrecht (p. 225234).

Walton, D. N. (1998). *The new dialectic*. University of Toronto Press.

Walton, D. N., & Krabbe, E. C. W. (1995). *Commitment in dialogue*. State University of New York Press.

Wells, S., Lozinski, P., & Pham, N. M. (2008). Towards an arguing agents competition: Architectural considerations. In Proceedings of 8th CMNA (computational models of natural argument) workshop, *European conference on artificial intelligence* (ECAI), university of Patras, Greece.

Wells, S., & Reed, C. (2004). Formal dialectic specification. In I. Rahwan, P. Moraitis, & C. Reed (Eds.), *First international workshop on argumentation in multi-agent systems*.

Wells, S., & Reed, C. (2012). A domain specific language for describing diverse systems of dialogue. *Journal of Applied Logic*, 10(4), 309–329.

Wooldridge, M. (2002). *An introduction to multiagent systems*. John Wiley & Sons, Ltd.

Yuan, T. (2004). *Human computer debate, a computational dialectics approach* (Unpublished doctoral dissertation). Leeds Metropolitan University.

Yuan, T., Moore, D., & Grierson, A. (2003). A conversational agent system as a test-bed to study the philosophical model DC. In *Proceedings of the 3rd workshop on computational models of natural argument* (CMNA'03).

Yuan, T., Moore, D., & Grierson, A. (2007). A human computer debating system and its dialogue strategies. *International Journal of Intelligent Systems*, 22(1), 133–156.

Yuan, T., Moore, D., & Grierson, A. (2008). A human-computer dialogue system for educational debate, a computational dialectics approach. *International Journal of Artificial Intelligence in Education*, 18(1), 3–26.

Yuan, T., Moore, D., Reed, C., Ravenscroft, A., & Maudet, N. (2011). Informal logic dialogue games in human-computer dialogue. *Knowledge Engineering Review*, 26(3), 159–174.

Yuan, T., & Schulze, J. (2008). Arg!draw: an argument graphs drawing tool. In *The second international conference on computational models of argument* (COMMA), Toulouse, France. (pp. 62–68).

Yuan, T., Schulze, J., Devereux, J., & Reed, C. (2008). Towards an arguing agents competition: Building on Argumento. In *Proceedings of 8th CMNA (computational models of natural argument) workshop*, European conference on artificial intelligence (ECAI), university of Patras, Greece.

Yuan, T., Svansson, V., Moore, D., & Grierson, A. (2007). A computer game for abstract argumentation. In *proceedings of IJCAI'2007 workshop on computational models of natural argument*, Hyderabad, India. (pp. 62–68).

Chapter 10

Arguing About Open Source Licensing Issues on the Web

Thomas F. Gordon

Fraunhofer FOKUS, Berlin, Germany, thomas.gordon@fokus.fraunhofer.de

Abstract. In the European MARKOS project we have developed an interactive web application, called the MARKOS License Analyzer, to help software developers to efficiently and cost effectively assess open source licensing issues and minimize their legal risks. The license analyzer is based on the Carneades Argumentation System and provides support for constructing, visualizing, evaluating and comparing competing legal arguments and theories. Arguments can be constructed both manually, via the Web interface, and automatically, from a knowledge-base consisting of an OWL ontology, a rulebase of argumentation schemes about open source software licensing issues, and facts in an RDF triplestore mined automatically from open source repositories such as GitHub.

1. Introduction

Copyright and licensing issues can cause important problems for developers of open source software, especially independent and part-time developers without access to a legal department or the financial resources to pay for legal services. These legal problems are dramatically exacerbated by the global nature of open source software. Despite the existence of international copyright treaties, copyright law is fundamentally national and a full analysis of copyright issues can thus require a deep knowledge of the copyright law of the countries of origin of all software used by a project, as well as an understanding of conflict of laws procedures and rules, to determine which laws apply to each issue.

Copyrights are monopolies created by law, granting copyright owners for a limited time certain exclusive rights to works for the public purpose of promoting science, learning and the arts. Ironically, the complexity and diversity of international copyright law creates business risks that can serve to inhibit innovation and creativity, also with regard to the development of open source software. The difficulty of determining licensing rights and obligations can have a chilling effect on the development and use of open source software, especially considering the high penalties for copyright infringement available in some jurisdictions.

In the European MARKOS project we have developed software tools to help software developers to efficiently and cost effectively assess open source licensing issues and minimize their legal risks. The diversity and complexity of copyright law makes this a challenging task. The idea of creating a fully automatic knowledge-based system for copyright law is doomed to fail, both because of the knowledge-acquisition bottleneck and due to the fundamental nature of legal reasoning, in which theories of the law and facts are constructed and compared by interpreting legal source texts and balancing evidence. Legal problems are in general ill-defined and not even semi-decidable.

Our basic idea for providing software tools which can help developers to effectively analyze licensing issues in a legally adequate manner, without placing unrealistic demands on their time or presuming too much prior knowledge of copyright law, takes a two-pronged approach:

1. A fully automatic *license checker*, developed by Massimilio di Penta and his colleagues at the University of Sannio, Italy, makes a first, rough analysis of licensing issues. The idea is to err on the safe side, by identifying and signaling potential copyright issues and explaining the causes of the issues.

2. If any issues are signaled by the license checker, the developer can choose to either resolve the cause of the issue, for example by choosing to use another license for his own software or another component with a compatible license, or to invest time in a more thorough investigation of the legal issues, using the *license analyzer* (LA), which is an interactive web-application, based on the Carneades argumentation system. The license analyzer provides support for constructing, visualizing, evaluating and comparing competing legal arguments and theories. Arguments can be constructed both manually, via the web interface, and automatically, from a knowledge-base consisting of an OWL ontology, a rulebase of argumentation schemes about open source software licensing issues, and facts in an RDF triplestore mined automatically from open source repositories such as GitHub. The argumentation schemes can be configured using *legal profiles* to express assumptions about how to apply general copyright concepts, such the concept of a 'derivative work', to software.

This paper presents an overview of the MARKOS license analyzer and is organized as follows. After introducing, in the next section, the MARKOS project as a whole, including its goals, partners and the system architecture and components of the MARKOS system, there is a section summarizing the software requirements identified in the project for the license analyzer. The requirements section emphasizes the complexity and openness of the problem, due to the international nature of software development, the jurisdictional and temporal variability of copyright law, and the 'open-texture' (Hart, 1961) of legal concepts, in particular the copyright concept of a 'derivative work'. Next, the three-tiered, web-based, system architecture of the license analyzer is explained, including how we have applied the Carneades argumentation system to meet the identified requirements. The Carneades inference engine is used to automatically construct arguments about licensing issues from an OWL ontology and rulebase of argumentation schemes, together with facts in the MARKOS triplestore. The system includes an OWL ontology and rulebase of argumentation schemes, presented in the following section, about open source licensing concepts and relations, focusing on license compatibility issues. Next, we provide a tour of the web user interface and demonstrate the main features of the system from the user's perspective. The final sections discuss related work and present our conclusions.

2. The MARKOS Project and System

MARKOS (The MARKet for Open Source) is a European research and development project (FP7-ICT-2011-8), which ran for 2.5 years from October, 2012, to March, 2015, with 8 partners from 5 countries, including companies, universities and research institutions (Bavota et al., 2014).[1] The main goal of the project is to develop an open source platform for the web, the MARKOS system, for providing a variety of innovative services to open source software developers:

- Semantic searching and browsing of software releases and other software entities available in open source repositories, such as SourceForge or GitHub, across the boundaries of projects, repositories, programming paradigms and languages, based on a semantic model of software concepts and relations. The semantic model of open source software, represented as an OWL ontology and stored in a RDF triplestore repository, provides an integrated view onto open source software available from repositories on the Internet, and enables code to be search and browsed at a high-level of abstraction.
- Identification of the provenance release of every open source software component, i.e. the original source code of the component, even when the source code has been copied, modified and reused.
- Identification of the open source license(s), such as the GNU Public License (GPL) or Apache license, applied to each software release, as well as the particular version of the licenses.

---

[1] http://www.markosproject.eu/

- Identification of dependencies and relationships among software releases and other software entities, such as the software libraries linked to, or the interfaces implemented.
- Assistance with the analysis of license compatibility issues, including help with the selection of open sources licenses which can be compatibility used by a software release without violating the terms and conditions of licenses of other software used, help with finding software components of a particular kind with licenses compatible with a given license, when used in a particular way, and, finally, help with analyzing and understanding possible license compatibility issues of a particular software release.

No prior system or service provides this functionality in its entirety. Integrated development environments (IDE) allow the structure of code to be browsed, but only after the software has been downloaded from a repository and loaded into the IDE. Source code repositories enable software in the repository to be searched, but not across the boundaries of repositories, and not in a deep, semantic way, but rather only via full-text search of the source code.

Figure 1 shows a high-level view of the MARKOS system architecture. Starting at the bottom, the *crawler* goes out and retrieves, at regular intervals, metadata about software releases and components, including their updates and modifications, from open source repositories on the Internet, such as GitHub or SourceForge. The crawler passes this metadata on to the *code analyzer* which retrieves and mines the full source code from the repositories to extract semantic information about software entities and their relations, using the MARKOS software ontology, and stores this information in the *semantic store,* an RDF triplestore. The information in the repository may be browsed and searched in two ways: 1) by other software on the Web, via the *linked data access point*, and 2) by end users, via the web *front end* (user interface). MARKOS also provides an *email notification service*, which users can subscribe to, to obtain news about updates to software entities, and a *correction manager* which allows project owners to manually edit, correct and complete the information automatically extracted by the crawler and code analyzer using heuristic methods. Finally, the *license assistant* interacts with the semantic store and users, via the front end, and consists of the fully automatic *license checker*, which is executed by the code analyzer, and the interactive *license analyzer*. It is the license analyzer which supports argumentation about licensing issues on the web and is the focus of the rest of this paper.

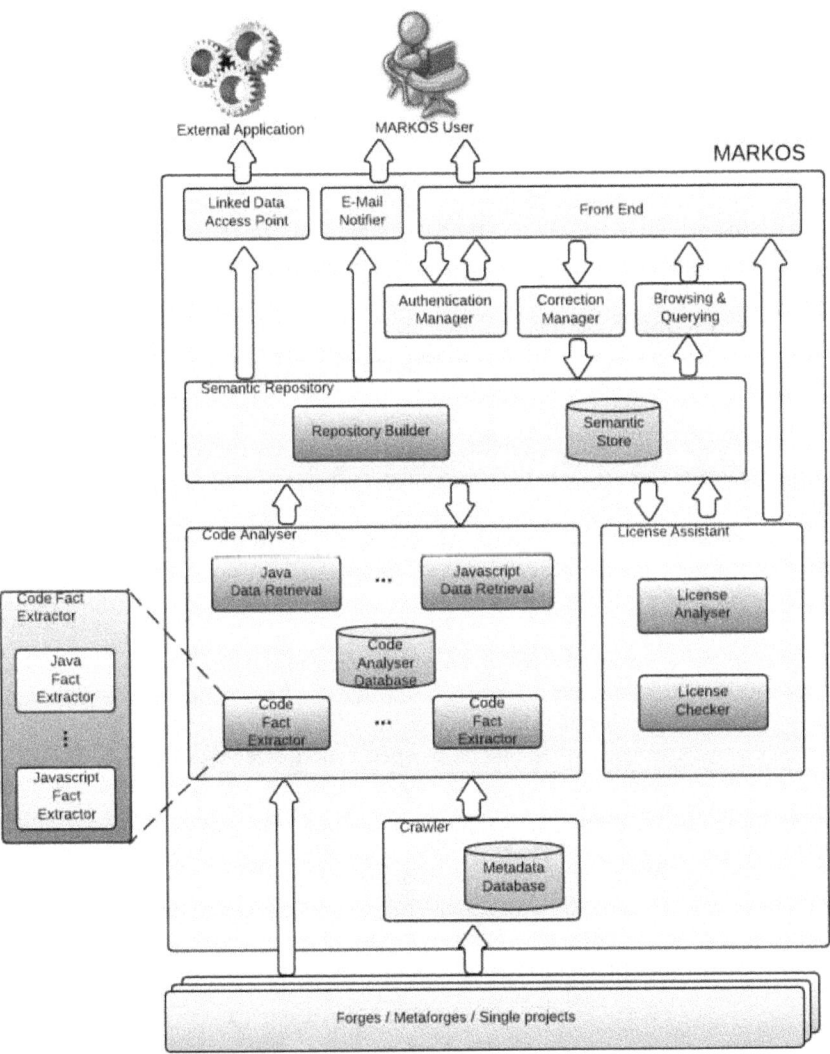

Figure 1. MARKOS System Architecture

3. License Analysis Requirements

A proper legal analysis of open source license compatibility issues cannot be performed automatically, in a mechanical way, but requires the interpretive and creative skills of a human to construct and compare theories of the facts and law, and relationships between them, in the light of the particular material facts of a case and the legal norms of the applicable jurisdiction. In legal practice there is never a uniquely right answer to some legal issue. Even if one takes the position that in principal there must be one right answer, in practice reasonable people can and will disagree about what this answer should be. Good arguments can always be made on both sides of any issue. Deciding legal issues requires good judgment, not just good logic. Legal problems are not well-formed and thus cannot be fully automated. Legal reasoning is a creative, synthetic process involving the construction, evaluation and comparison of theories. While formal, analytical methods can be useful for analyzing the logical consequences of these theories, no formal method can generate all possible theories, since the search space of theories is not enumerable. This nature of legal reasoning leads to some necessary uncertainty and risk which cannot be entirely eliminated. This is as true for open source software development as for any other activity regulated by law.

There are various ways to attack arguments: by attacking a premise, by putting forward an argument, called a rebuttal, for a contrary conclusion, or by undercutting the argument with an argument claiming that its major premise does not apply in this case. The process of making claims, putting forward arguments and deciding issues is regulated by rules of procedure. These procedural rules regulate, among other things, the distribution of the burden of proof among the parties and the proof standard, such as the civil law preponderance of evidence standard for resolving issues.

Open source software developers are interested in *avoiding* licensing issues and legal conflicts, not just in *resolving* legal conflicts and protecting their rights, after a conflict arises. The essential undecidability of legal reasoning creates practical problems and legal risks when selecting open source licenses. These problems and risks are compounded by the international character of open source software development. A thorough analysis of the licensing issues would require not only an understanding of the copyright law of one's own country, but also the copyright law of the other countries where software used by the project was developed or published, in additional to relevant international treaties. It may not always be feasible to perform a thorough legal analysis. Open source developers often are volunteers working unpaid in their free time and cannot afford to hire lawyers to professionally assess licensing issues.

These considerations led to the requirement of developing an effective decision procedure for applying simplified, de facto norms, represented in a rulebase, to construct arguments about whether or not some open source license may be used in some software project without violating the licensing conditions of other software used by the project. This approach does not suffer from the failings of *mechanical jurisprudence*, which has been deemed wholly inadequate from the perspective of legal theory, since the rules are used only to help the user to find and construct arguments, in a transparent way, without automatically deciding the legal issues. The resulting arguments, visualized in an argument map, can help the user to make sense of the issues and ask critical questions about the arguments, including challenging the simplifying assumptions made when developing the rulebase.

Arguing about open source licensing issues on the Web

Here is a brief *scenario* illustrating how the MARKOS License Analyzer is expected to be used by open source software developers. A scenario is a fictional story, illustrating how actors in various roles interact with the system to be developed in order to achieve some goals or produce some valuable result. Scenarios are not intended to be comprehensive, illustrating all the features of the system, but rather only to illustrate the main use cases.

1.1 Application Scenario

Silke is a software architect responsible for the license quality of C1, a software house specialized in the development of open source software. She needs to check if the P1 software is violating any license.

Usually Silke works with the technical lead and the legal office of her company to evaluate the legal risks, but the P1 software has become so large, with many dependencies on other components, that it is becoming increasingly difficult for them to evaluate the risks comprehensively. After a while she learned about MARKOS from a blog.

Silke searches for and finds P1 on MARKOS and goes to the 'license analysis' section. She notices the warning on the license analysis page, suggesting that the L1 license used by P1 may be incompatible with the L2 license of the software entities used by P1. This warning was produced automatically by the MARKOS license checker after the P1 software was indexed by the MARKOS code analyzer.

Silke wants to analyze the license compatibility issues more thoroughly so she clicks the 'Analyze' button. In an interactive dialogue with MARKOS, Silke enters further facts about the P1 software, if necessary, and obtains an argument graph showing dependencies between various legal assumptions and conclusions about license compatibility and other legal issues. Silke selects a legal profile containing particular assumptions about the governing copyright law to have the argument graph evaluated with these assumptions.

Silke's analysis of P1 has helped to clarify the potential legal risk. The P1 software uses a L1 license, but is dynamically linked to an P2 component that uses a L2 license. The L1 license is not compatible with the L2 license. This might not be a problem if dynamic linking does not cause the P1 software to be derived from P2, but Silke has selected a legal profile which follows the legal opinion of the Free Software Foundation, which considers all forms of linking to create derivative works.

A link to the argument graph can be published and stored in the MARKOS repository. The argument graphs for the analyses which have been carried out for a software entity are listed on the project page. Silke wants to keep her analysis of the license issue of the P1 software private for the time being, at least until the issues have been resolved, and thus decides not to publish the link.

Silke discusses the problem with the technical lead. He uses the MARKOS site to find the license of P2 and sees that it is part of an L2 project. His team had copied just the object code of P2 from another project that, perhaps erroneously, also used the L1 license. But a thorough analysis of the other project would be required to determine if their use of L1 was incorrect, since this would depend on how they used P2. Merely including a copy of the object code of P2 with their program probably would not require them to use the L1 license for their own work.

Silke discusses the legal issues with the legal office of her company. The legal office is of the opinion that dynamic linking does not create works according to the law they believe would govern the case. With the aid of the legal office, Silke creates a new legal profile tailored to their view of the governing law. Silke then performs a new analysis of the licensing issues, using this new profile.

The new analysis, using the revised legal profile, provides arguments justifying the publication of the P1 software using an L1 license, as desired by Silke's company. Nonetheless, Silke decides against publishing this analysis on the MARKOS server. The legal department of her company recommend 'letting sleeping dogs lie' by keeping the analysis private until someone, such as the owner of the P2 software, makes an issue out of the use of the L1 license.

To be on the safe side, the technical lead searches MARKOS for another K1 component, using user-defined keywords, but with a L2 license. MARKOS was able to find a couple of possible alternative components, but a closer evaluation of the alternatives lead C1 to accept the legal risks of continuing to use P2, which they consider to be best from a technical perspective.

User Stories

Following the methodology of agile software requirements (Leffingwell, 2010), the functional requirements of the license analyzer have been specified as 'user stories': brief, high-level statements describing users in particular roles (who) would like to be able to use the system to perform some task (what) in order to achieve some benefit or value (why).

Five user roles were identified: copyright lawyer, legal knowledge engineer, software architect, and system administrator. For simplicity and to save space, we do not distinguish these roles in this paper.

The identified requirements have been grouped into three categories (management of legal profiles, license assessment, and copyright law modeling), as follows. Only the main requirements are shown.

Management of Legal Profiles

We use the term *legal profile* in MARKOS to mean a set of assumptions about the valid legal norms of some jurisdiction.

Legal Profile Selection
Choosing a legal profile to use to analyze license compatibility issues, reflecting the governing law of a particular jurisdiction or interpretation of copyright concepts.

Legal Profile Definition
Defining legal profiles matching some interpretation of copyright concepts or the law of some jurisdiction, to make them available for use when analysing license compatibility issues.

License Assessment

License Checker
A fully automatic 'quick and dirty' license compatibility check, to be used as a heuristic for finding issues to analyze more thoroughly.

License Analysis
Interactively construct, evaluate and visualize arguments pro and con license compatibility issues, using a copyright ontology and rulebase, along with facts from the semantic repository, to facilitate a deeper understanding of the issues, helping to avoid legal risks and choose appropriate licenses.

Find Compatible Open Source Licenses
Search automatically for open source licenses compatible with the licenses of all software releases used by a given software release, according to a selected legal profile, to help users to choose compatible licenses and reduce legal risks.

Find Software Entities with Compatible Licenses
Search for software entities with selected tags which have licenses which are presumably compatible with a particular license, given a specific way the entity will be used and a particular legal profile, so that users can focus their evaluation of software entities on those with compatible licenses.

The next three requirements are about visualizing, editing and publishing *argument graphs*. Here, an argument graph is a directed, bipartite graph of statement nodes and argument nodes, which represents dependencies and relations between the premises and conclusions of arguments, and an *argumentation scheme* is a template or form which can be used to construct arguments of a particular type, by instantiating the scheme, i.e. by filling in the blanks in the template.

Argument Graph Visualization
View and browse maps of argument graphs, with different views focusing on different parts of the graph, and at different levels of abstraction, to promote a better understanding of the reasoning behind the analysis.

Argument Graph Editing
Use argumentation schemes to edit argument graphs, to compensate for the possible incorrectness or incompleteness of the facts in the semantic repository and the ontology and rulebase modeling copyright law.

Publish License Analysis
Publish a link to an argument graph with the results on an analysis of license compatibility issues in the MARKOS repository, to facilitate the sharing of license analyses with other developers.

Copyright Law Modeling

The final three requirements are about supporting knowledge engineers with the task of developing and maintaining an ontology and rulebase modeling copyright law for the purpose of analyzing open source software license compatibility issues.

Ontology
Define an ontology of concepts and relations for software licenses and copyright concepts, including relationships between software works, such as whether one work has been derived from another, for use in argumentation schemes for reasoning about open source license compatibility issues.

Rules and Argumentation Schemes
Use a rule language to define strict and defeasible argumentation schemes for reasoning about open source software license compatibility issues, for use by an inference engine to automatically construct arguments from facts in the semantic repository.

Modeling of License Templates
Use the ontology to formally model the terms and conditions of common open source license templates, such as the GNU Public License (GPL) or the Apache License, to enable the analysis of license compatibility issues.

4. System Architecture

Figure 2 is a sequence diagram showing interactions between the license analyzer, front end, semantic store and repository of the MARKOS system. The user initiates the license analysis by clicking on a link in the front end. This sends an analyze command via HTTP to the server-side of the license analyzer and returns its Web client, implemented as a Rich Internet Application (RIA) in HTML, JSON and CoffeeScript, using the AngularJS and Bootstrap JavaScript libraries. The Web client is opened in a new tab of the user's web browser. The server-side of the license analyzer retrieves facts about the software release to be analyzed from the

Arguing about open source licensing issues on the Web

MARKOS semantic store, via SPARQL Protocol and RDF Query Language (SPARQL) queries. This includes information about other software releases used, together with their licenses. The results of the analysis are visualized both as hypertext, in HTML, and an interactive argument map, implemented using the Structured Vector Graphics (SVG) language. The user remains in control of the analysis process and can use a forms-based argument editor, with customized forms controlled by argumentation schemes, to enter new arguments or modify existing arguments, including those generated automatically by applying argumentations schemes in the copyright rulebase. The resulting argument map is stored in a relational database on the server-side of the license analyzer. Finally, the user can optionally publish the license analysis, by first clicking on a button and then completing and submitting a form, which causes a URI referencing the argument map to be stored in the MARKOS semantic repository, along with further information, such as the userid and email address of the person who performed the analysis. The front end lists the published argument maps for previous analyses. Clicking on a link in the list opens the web client of the license analyzer and displays the selected argument map.

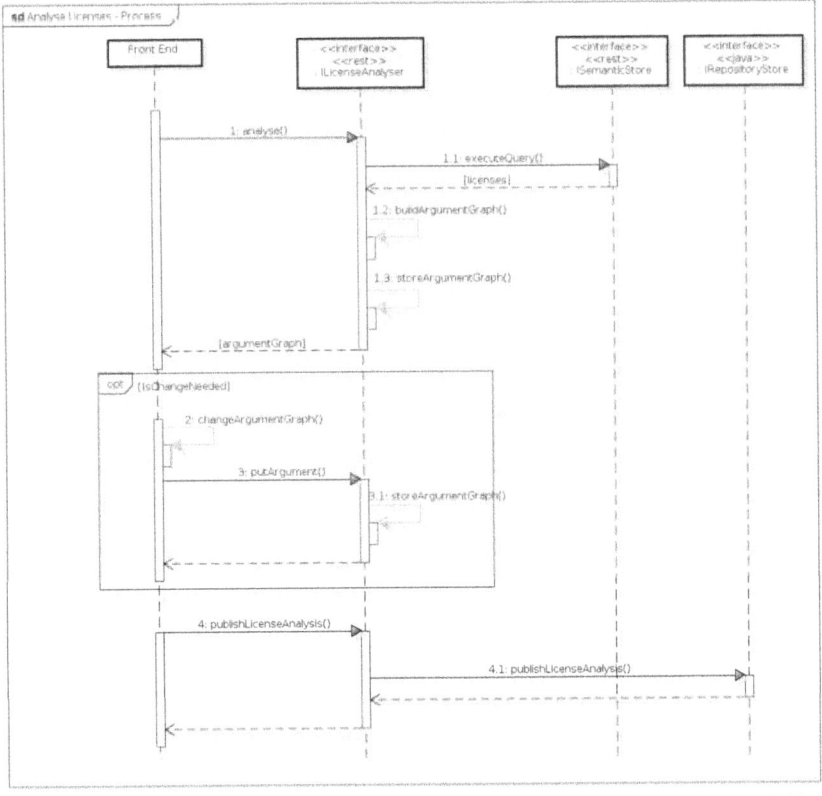

Figure 2. License Analysis Sequence Diagram

The server-side of the license analyzer is based on Version 3.7 of the Carneades argumentation system. This version of Carneades is a three-tiered web application, with a relational database backend. The application tier of the system is implemented in Clojure, a Lisp dialect which compiles to bytecodes of the Java Virtual Machine (JVM). The server communicates with the Web client via RESTful web services, with messages encoded in JSON.

The application tier of Carneades consists of several modules (Clojure packages). The main modules are:

Dung
An implementation of grounded semantics for Dung abstract argumentation frameworks (Dung, 1995).

CAES
An evaluator for the original Carneades model of structured argumentation (Gordon, Prakken, & Walton, 2007), which supports variable proof standards, such as preponderance of the evidence and beyond reasonable doubt, and the distribution of the burden of proof by distinguishing two kinds of critical questions of argumentation schemes, modeled as exceptions and assumptions.

ASPIC
An evaluator for a model of structured argumentation based on ASPIC+ (Prakken, 2010), which is capable of handling cyclic argument graphs via a mapping to Dung abstract argumentation frameworks.

Theory
A module for representing domain knowledge in a rule language for argumentation schemes, embedded in Clojure. The language can import OWL ontologies. Argumentation schemes can be organized in a hierarchical structure, similar to the chapters, sections and subsections of a book. The theory as a whole and each section can be annotated with metadata, including textual descriptions in multiple natural languages, including markup in the Markdown language. These features are designed to support the *isomorphic* modeling of legislation in legal applications (Bench-Capon & Coenen, 1992; Bench-Capon & Gordon, 2009). Emdedding the theory language in Clojure enables theories to import other theories, using the Clojure packages in the usual way.

Search
A module for representing, traversing and searching (possibly) infinite search spaces. Currently only depth-first and breadth-first search methods are provided. To assure termination, a resource bound limiting the number of nodes expanded in the search space can be provided. Lazy sequences are used to expand nodes as they are visited.

Argument Generator
This module defines a generic protocol for generating arguments, allowing multiple methods for constructing arguments to be used together. For example by asking the user, retrieving information from a relational database, querying a semantic repository using SPARQL, or invoking an external inference engine. The main challenge is to find a way to extract arguments from the source.

Ask
An argument generator which asks the user for information, in the manner of the dialogue component of classical expert systems. The questions are presented in natural language, using forms (templates) defined for the concepts and relations used. The forms are specified in rulebases, called 'theories' in Carneades.

Argument Construction
A generator for arguments from theories. This is a backwards chaining inference engine which applies argumentation schemes defined in theories to statements in an argument graph. The inference engine uses resource-bounded depth-first search and is thus guaranteed to terminate. All arguments found are returned, even if the goal was not satisified. The user can restart the search for any unproven premises, to continue the process of trying to prove the main goal. The inference engine uses tabling to assure that no argumentation scheme is applied more than once to any goal.

Triplestore
This is a module for constructing arguments from facts stored in a triplestore. The triplestore is queried using SPARQL. This module is used by the MARKOS license analyzer to retrieve facts from the MARKOS semantic repository.

LACIJ
This module generates visualizations of argument graphs, called *argument maps*, using the Scable Vector Graphics (SVG) format, which is a World Wide Web Consortium (W3C) standard supported by modern Web browsers. The automatic graph layout algorithms were implemented by Pierre Allix in Clojure, one of the lead developers of Carneades. The code has been published as a separate open source library on Github.[2]

Serialization
This module provides export and import procedures for AIF, the Argument Interchange Format (Chesnevar et al., 2006), serialized using JSON, and our own Carneades Argument Format (CAF), using XML. Exporting to CAF is lossless, since the format was designed especially for Carneades argument graphs.

---

[2] https://github.com/pallix/lacij

5. Ontology and Rulebase

The MARKOS license analyzer applies an inference engine to a knowledge-base about open source license compatibility issues to assist users with constructing arguments about licensing issues. The knowledge base consists of an OWL ontology, specifying the domain language of predicates (classes, properties) and terms (individuals) needed, together with a rulebase (theory) of domain-specific argumentation schemes using these predicates and terms.

The ontology consists of several parts, including a software ontology, based in part on the Unified Modeling Language (Fowler & Scott, 2000), a high-level copyright ontology and, building upon these, a more specific ontology modeling open source software licenses.

The copyright ontology includes classes for various kinds of rights (economic, exclusive, fair use, and moral rights) and works (collections, original works and protected works), as well as a fairly detailed and comprehensive model of licensing terms (e.g. adaptation, attribution, distribution, notices, patent grants, reciprocity, publication of source), based on ccREL, the Creative Commons Rights Expression Language (Abelson, Adida, Linksvayer, & Yergler, 2008).

The open source licensing ontology models the most widely used open source software licenses (more precisely license *templates*), including AGPL, Apache, Artistic, GPL, BSD, EPL, EUPL, LGPL, MIT and MPL, some of these in multiple versions.

OWL ontologies, being based on classical logic, are not sufficient, in general, for modeling legal norms, for many reasons (Gordon, Governatori, & Rotolo, 2009). There can be multiple, competing interpretations of legal source texts, such as legislation and court decisions. Rules from different levels of a legal system, such as constitutional, federal and state law, can conflict. Laws change over time, so newer rules can conflict with older rules. Meta-level principals exist to resolve conflicts among conflicting interpretations of the law, or conflicting legal norms from different authorities or over time. Some way is needed to 'reify' legal rules and their properties, to be able to reason about and resolve rule conflicts.

We use argumentation schemes to overcome one of the main difficulties in developing a knowledge-base to help developers to resolve open source license compatibility issues: open source software is shared and developed world-wide on the Internet, but there is no universal copyright law. Despite efforts to harmonize copyright law, differences among legal systems remain. For example, whereas the European Court of Justice (ECJ) decided in SAS Institute Inc. v World Programming Ltd., [2013] EWHC 69 (Ch), that Application Programmer Interfaces (APIs) are not protected by copyright, a Federal Circuit court in the United States decided that APIs are indeed protected by copyright. When there are multiple, conflicting norms, or multiple interpretations of a norm, as in this example, we include multiple argumentation schemes in the model, one for each of the competing norms or interpretations and provide a way for users to choose among them. Users can define 'legal profiles' to configure the knowledge-base to match their own preferred interpretation of copyright law, or the governing law of relevant jurisdictions. The system is flexible enough to enable different rules to be applied to different issues of the same case. This can be important, as the governing law may be different for different software components of a project.

The preferred approach in the Artificial Intelligence and Law community for modeling legislation is to structure the model in a way which matches the structure of the legislation (Bench-Capon & Coenen, 1992; Bench-Capon & Gordon, 2009). Unfortunately, this approach is not feasible for a copyright law knowledge-base for resolving open source license compatibility issues, since it would require a separate model for each country, along with the corresponding depth and breadth of knowledge of the language and law of every country. Thus we have adopted a more practical approach for this application. The knowledge-base is a reconstruction of the *de facto* norms of copyright law, as they are understood and applied by open source software developers in practice, world-wide, with multiple rules for controversial issues. The model includes a facility for describing the rules in multiple natural languages and for linking the rules in these descriptions to the relevant legal codes of national copyright legislation.

Here is an overview of the classes defined in the ontology for copyright license templates:

- CopyrightLicenseTemplate
    - OpenSourceLicenseTemplate
        - AcademicLicenseTemplate
        - ReciprocalLicenseTemplate
            - StrongReciprocalLicenseTemplate
            - WeakReciprocalLicenseTemplate

We call these license *templates*, rather than licenses because, when a software developer applies, for example, the GPL to a software release, the developer is granting a particular license, an *instance* of the GPL template, to users of the software. The developer, not the Free Software Foundation, is the owner of the software and the *licensor* of the license. The users are the *licensees* of the license.

Below is an example showing how a license template, here Version 2 of the Mozilla Public License (MPL-2.0), is modelled in the OWL ontology using our reconstruction in OWL of the Creative Commons Rights Expression Language:

```
Individual: MPL-2.0

    Annotations:
        rdfs:seeAlso
"http://www.mozilla.org/MPL/2.0/"^^xsd:anyURI,
        mtop:name "MPL"

    Types:
        WeakReciprocalLicenseTemplate
```

```
Facts:
  copyright:permits copyright:Sublicensing,
  copyright:permits copyright:Reproduction,
  copyright:permits copyright:PatentGrant,
  copyright:permits copyright:Adaptation,
  copyright:permits copyright:Distribution,
  copyright:requires copyright:SourceCode,
  copyright:requires copyright:Notice,
  copyright:prohibits copyright:HoldLiable,
  copyright:requires copyright:Reciprocity,
  copyright:prohibits copyright:UseTrademark,
  mtop:version     "2.0",
```

Again, the ontology specifies the language of predicates and terms used by theories of legal norms regarding open source license compatibility issues, formalized using the Carneades rule language for argumentation schemes. The Carneades rule language is based on Horn clause logic, with extensions for defeasible reasoning. In classical logic, Horn clauses are interpreted as material conditionals and a single inference rule, resolution, is used to derive strict (non-defeasible) inferences. In Carneades, we do not interpret rules as material conditionals, but as domain-dependent defeasible inference rules, called 'argumentation schemes' (Walton, 1996). Instantiating a rule constructs an argument which may be attacked by other arguments in various ways.

Argumentation schemes are represented using a domain specific language embedded as symbolic expressions in the Clojure programming language. This enables theories to be defined in Clojure packages and to take full advantage of the expressiveness of the Clojure language and development tools. For example, rules can include premises which apply Clojure functions to compute values, comparable to 'procedural attachment' in expert systems using production rules.

Carneades includes representations of many of Walton's argumentation schemes (Walton, Reed, & Macagno, 2008), such as argument from expert opinion. But here we would like to show how domain specific argumentation schemes, in this case for open source license compatibility issues, have been modeled. Two example schemes are shown below. The first expresses the default licensing rule, stating that in general any license template may be applied to any work. The second scheme expresses an exception to this general rules, for works derived from works licensed using reciprocal (copyleft) licenses, like the GPL, *unless* the licenses are compatible.

```
(t/make-scheme
  :id 'default-licensing-rule
  :weight 0.25
  :header (dc/make-metadata
            :title "Default licensing"
            :description {:en "Presumably, a work may be
                licensed using any license template."})
  :conclusion '(copyright:mayBeLicensedUsing ?W ?T)
  :premises [(a/pm '(lic:CopyrightLicenseTemplate ?T))])
```

Arguing about open source licensing issues on the Web

```
    (t/make-scheme
       :id 'reciprocity-rule*
       :header (dc/make-metadata
                :title "Reciprocity"
                :description {:en "A work W1 may not use a
license
                   template T1 if the work is derived from a work
W2
                   licensed using a reciprocal license template
T2,
                   unless T1 is compatible with T2."})
       :pro false
       :conclusion '(copyright:mayBeLicensedUsing ?W1 ?T1)
       :premises [(a/pm '(lic:CopyrightLicenseTemplate ?T1))
                  (a/pm '(copyright:derivedFrom ?W1 ?W2))
                  (a/pm '(lic:licenseTemplate ?W2 ?T2))
                  (a/pm '(ReciprocalLicenseTemplate ?T2))]
       :exceptions [(a/pm '(copyright:compatibleWith ?T1
?T2))])
```

These schemes illustrate two ways schemes can be applied to construct attacking arguments:
1. The second scheme can be used to construct *rebuttals* to arguments constructed using the first scheme, because the second scheme constructs arguments *con* the conclusion of the first scheme.
   The exception for compatible licenses of the second scheme can be used to construct an argument *undercutting* the argument constructed using the scheme.

6. User Interface

The user interface of the MARKOS license analyzer is an interactive (rich) Web client created by customizing the Carneades Web client, using CSS style files, so that it matches the look and feel of the rest of the MARKOS front end. Figure 3 shows a view of a project release in the MARKOS front end.

The content explorer panel, on the bottom left, shows the components of the software release and indicates whether they have been defined in this project or imported from other projects. That is, one can easily see whether the *provenance release* of the component is from this project or some other project.

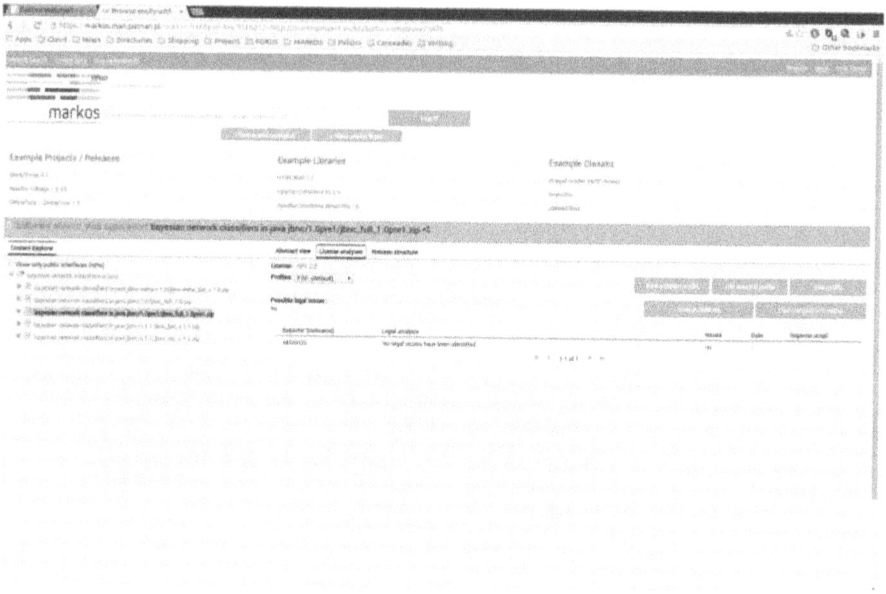

Figure 3. View of a Project Release in MARKOS

The panel to the right includes several tabs for viewing further information about the release. In the figure, the license analysis tab has been selected. It shows the license (template) identified by the MARKOS code analyzer, in this case GPL 2.0, and the results of the automatic license checker. In the example shown, the license checker found no licensing issues. At the bottom there is a table showing the license analyses which have been published. The first item shows (again) the results of the license checker. The other items references argument graphs constructed using the license analyzer which have been published. In the example, none of been published yet.

Whether or not the license checker has identified an issue, the license analyzer can be used to double check the results and obtain a better understanding of the issues, or why there is no issue. To start the license analyzer, the user can first select one of the existing legal profiles from the pull-down menu. In the figure, the Free Software Foundation (FSF) profile has been selected. Following the interpretation of copyright law of the FSF, this profile includes a rule stating that linking to a software library creates a derivative work. There are buttons to the right for viewing or editing the selected profile, or to create a new profile based on the selected profile.

Clicking on the 'Analyze licenses' button starts the license analyzer using the selected profile. This starts the Carneades inference engine, applying the rules of the license compatibility theory to the facts about this project release and its dependencies, retrieved from the MARKOS semantic store, and displays an outline of the resulting argument graph, as shown in Figure 4.

Arguing about open source licensing issues on the Web

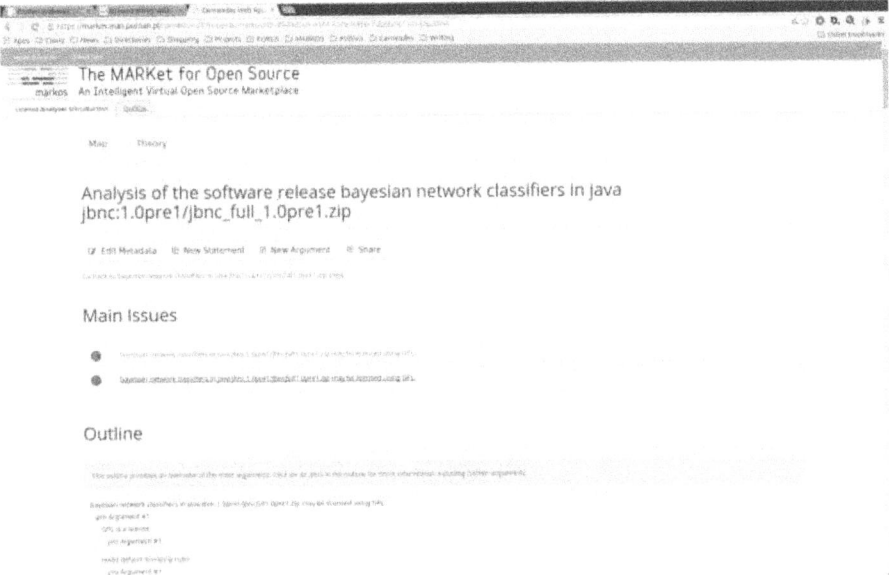

Figure 4. Argument Graph Outline

The argument graph may also be viewed in an argument map (diagram), by clicking on the 'Map' button at the top of the outline view, as shown in Figure 5.

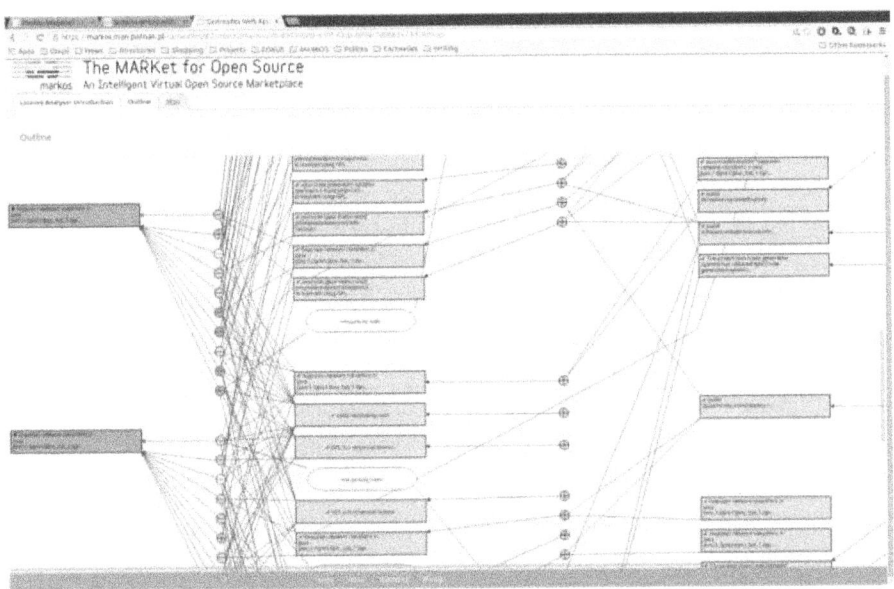

Figure 5. Argument Map

The argument and statement nodes in the argument contain hyperlinks. Clicking on a node shows more detailed information. For example, Figure 6 shows the details of an argument node.

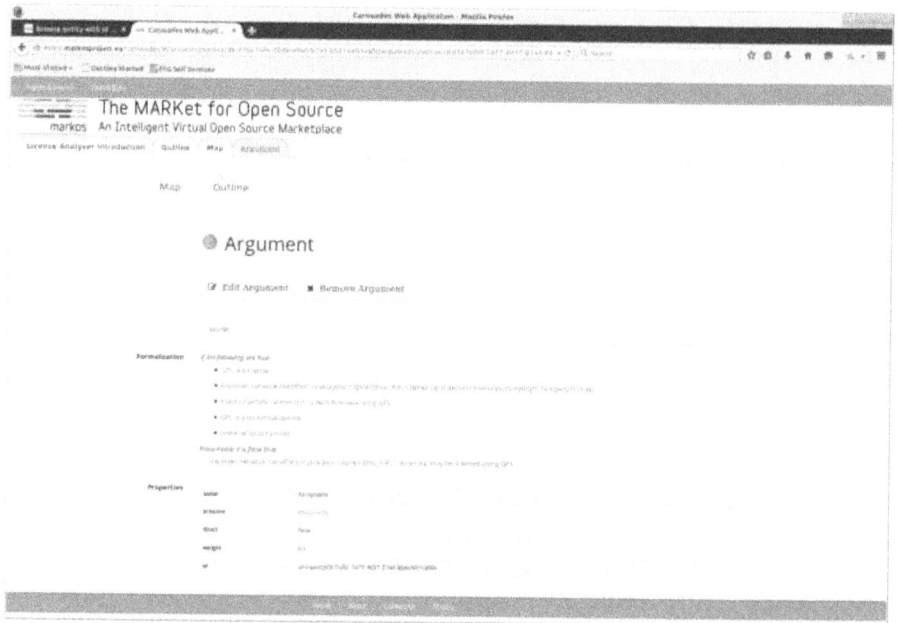

Figure 6. Details of an Argument Node

The Argument can be edited, guided by argumentations schemes, by clicking on the 'Edit Argument' button in the view of the argument node, as shown in Figure 7.

Arguing about open source licensing issues on the Web

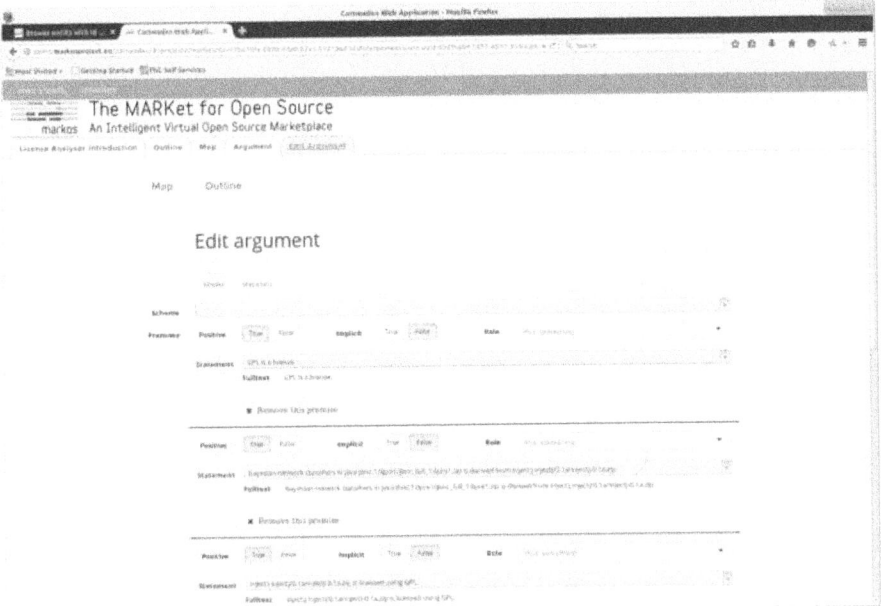

Figure 7. Argument Editor

When the user is satisfied with the revised argument graph, it can be published and shared with others on the Web, by clicking the 'Share' button in the outline view, as shown in Figure 8. Currently the argument map can be published to the MARKOS server, by storing a link to the argument map in the MARKOS semantic repository, and posted on Twitter.

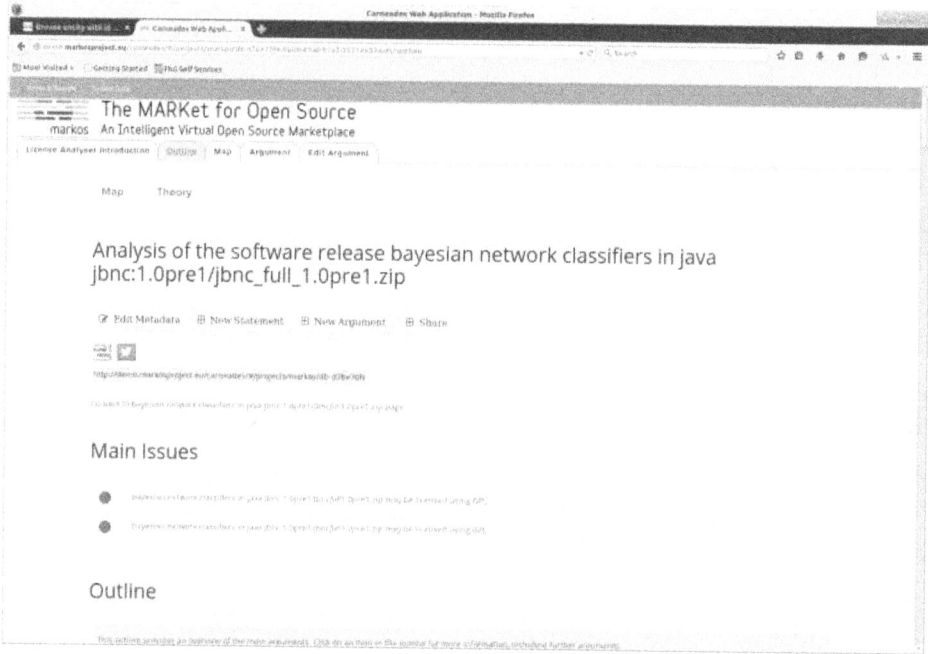

Figure 8. Publishing Argument Maps to MARKOS and Twitter

Related Work

Here we overview prior, related work on software for supporting the analysis of license compatibility issues and web-based tools for the construction, evaluation and visualization of arguments.

Let us begin with an overview of prior license analysis software. The GPL Compliance Toolkit (Coleman, 2001) included a 'License Report Wizard' which generates a report of all the licenses of executables and libraries used by a project and highlights which licences are incompatible with the license chosen for the project. LIDESC, the Librock License Awareness System (LIDESC, 2001), identified the license used by a software package and detects and reports licensing conflicts, based on an encoding of the terms and conditions of a number of open source licenses. OSLC, the Open Source License Checker (OSLC, 2007) analyses the source code of software packages to identify the licenses used by each package. These tools are all applications which are installed and run locally on the user's computer. Black Duck[3] and Palamida[4] provide online, web-based licenses analysis services. FOSSology is a open source application which provides similar

---

[3] https://www.blackducksoftware.com/
[4] http://www.palamida.com/

functionality.[5] ASLA, the Automated Software Analyzer (Tuunanen, Koskinen, & Kärkkäinen, 2009), is another tool which identifies licenses by analysing source code and checks for license compatibility issues. (Alspaugh, Asuncion, & Scacchi, 2009; Alspaugh, Scacchi, & Asuncion, 2010) presents a metamodel for software licenses, based on Hohfeldian (Hohfeld, 1913) legal concepts (e.g. rights and duties), and an algorithm based on this model for computing licensing conflicts.

These tools are all similar to the license checker of the MARKOS system, which is based on prior work of Massimiliano di Penta, one of the participants in the MARKOS project (D. M. Germán & Di Penta, 2012; D. M. Germán D. M.n, Di Penta, & Davies, 2010). They are focused on automatically identifying the licenses of software used by some package and applying a license compatibility model to detect and report potential licensing issues, without user intervention. They differ in the heuristics used to identify licences and dependencies among software releases.

The MARKOS license analyser presented here is a collaborative web-based application based on an initial single-user protoype developed in the European Qualipso project (Gordon, 2011) and improves upon these prior systems in the following ways:

- Legal profiles provide end users with a simple way to configure the model of copyright law and open source license compatibility issues to match a particular interpretation of copyright concepts, such as 'derivative work', or the law of a particular jurisdiction.
- Conflicting arguments, from competing rules, can be generated by the model, based on a defeasible logic of the kind used in practice by lawyers.
- Explanations of the analyses, in the form of argument graphs, are generated, along with several ways of viewing and browsing the argument graphs, using hypertext and argument maps.
- An argument editor is provided, enabling users to modify the arguments to compensate for limitations of the model and produce an analysis which reflects their own understanding of the law and facts.
- Analyses of licensing issues, represented as argument graphs, can be shared with other users on the web, via Twitter and the MARKOS platform.

On the other hand, the MARKOS license analyser does not itself identify the licenses used or dependencies among software releases. These tasks are performed by other components of the MARKOS system, in particular the code analyser and license checker.

Regarding related work on software applications for supporting argumentation, let us focus on Web applications. Rationale is a well-known and mature commercial, closed-source argument mapping sytem, supporting both 'standard' Beardsley/Freeman (Beardsley, 1950; Freeman, 1991) and IBIS (Kunz & Rittel,

---

[5] http://www.fossology.org/projects/fossology

1970) style argument graphs. Rationale used to be a Windows-only desktop application, but the most recent version is available as a Web service.[6]

AGORA-net is an interactive, collaborative and web-based argument mapping tool (Hoffmann & Borenstein, 2014). ARGORA-net by design supports only a small number (4) of argumentation schemes, all of which are deductively valid.

ArguNet consists of two parts, a desktop application, written in Java, for editing argument maps, and a separate web application, written in JavaScript, for browsing and viewing argument maps created with the editor (Betz & Cacean, 2011). ArguNet provides an argument database for storing and sharing argument maps on the Web. The argument maps are created and uploaded to the server using the ArguNet editor and viewed on the Web using the ArguNet browser Web application. ArguNet's model of argument structure is a hybrid of abstract and standard models. Attack and support relations between abstract arguments can be represented. Optionally, the premises and conclusions of individual arguments can be represented.

The Centre for Argument Technology (ARG-tech) in the School of Computing at the University of Dundee, Scotland, headed by Chris Reed, has developed a number of argumentation tools, including:

Argument Interchange Format (AIF)
An abstract data model for argument graphs, along with a number of concrete data formats implementing the abstact model using XML, RDF, OWL and JSON.

AIFdb
A web service providing access to a database of arguments representing using the data model of the Argument Interchange Format (AIF)

OVA (Online Visualization of Argument)
A web application for reconstructing arguments in web pages and saving them to the Argument Interchange Format Database (AIFdb)

OVAview
A Web application for viewing argument maps of arguments in the AIFdb database

The Dialogue Game Description Language (DDGL+)
A programming language for specifying argumentation protocols

The Dialogue Game Execution Platform
A Web service for conducting dialogues according to protocols defined using DDGL+.

Arvina
A Web application for engaging in dialogues, using the Dialogue Game Execution Platform, with software agents. The software agents put forward arguments available in the AIFdb database.

---

[6] https://www.rationaleonline.com

TOAST
A tool for evaluating AIF arguments using Dung abstract argumentation frameworks by mapping AIF to ASPIC+ (Prakken, 2010).

The ARG-tech center has produced other systems as well, including the widely-used Aracuaria desktop application for argument reconstruction and mapping. These other systems have either been superceded, like Aracauria, by the tools listed above, or are separate tools which are research prototypes for particular tasks which do not appear to be well integrated into the above set of tools at this time.

Deliberatorium is another Web application for collaboratively constructing and viewing IBIS argument maps (Klein, 2011). Neither the source code nor the object code of the system is available. Argument maps are displayed in hypertext outlines. No two-dimensional diagrams are available.

The LASAD (Learning to Argue: Generalized Support Across Domains) argumentation support system was developed to support online argumentation, with a focus on educational, learning and tutoring application scenarios (Loll & Pinkwart, 2013). The data model of LSAD is designed to be configurable to support a wide variety of argument models. An LSAD model is a directed graph with typed (labeled) nodes and edges (links). LSAD provides a way to define 'ontologies' to configure the system for a particular model of argument, where an ontology consists of the node and edge types, along with information about how to visualize the nodes and edges in diagrams, by specifing such properties as colors, border styles, arrowheads, and line thickness. Thus it easy to configure LSAD to support the IBIS or Toulmin models of argument. More complex argument data models, such as the ASPIC+ and Carneades models of structured argument, can be pressed into this generic mold, but the resulting editing and visualization tools do not have the usability or design quality that would be possible with custom users interfaces developed specifically for these data models. Moreover, LSAD provides no support for applying argumentation schemes or formal semantics to evaluate arguments.

To our knowledge, Carneades is currently the only open source, web-based argumentation system to support the automatic construction of arguments from a knowledge-base of ontologies and argumentation schemes, evaluated using state-of-the-art computational models of (defeasible) argument, with well-integrated tools for editing and visualizing argument maps.

7. Conclusions

In this paper, we have presented an application of the Carneades argumentation system for helping software developers to construct, evaluate, edit, visualize and share arguments about open source software licensing issues on the Web. Arguments can be constructed automatically, from an ontology and rulebase of argumentation schemes modeling relevant copyright law concepts and edited manually, using a scheme-driven Web user interface. The argumentation schemes can be configured at run-time, using profiles, to express assumptions about how to apply general copyright concepts, such as the concept of a 'derivative work', to software. The system is built upon a number of Semantic Web standards, including OWL ontologies, RDF triplestores and the SPARQL query language.

Acknowledgements

This work was supported by the European MARKOS project (FP7-ICT-2011-8).

References

Abelson, H., Adida, B., Linksvayer, M., & Yergler, N. (2008). CcREL: The creative commons rights expression language. Retrieved from https://wiki.creativecommons.org/images/d/d6/Ccrel-1.0.pdf

Alspaugh, T. A., Asuncion, H. U., & Scacchi, W. (2009). Intellectual property rights requirements for heterogeneously-licensed systems. In Requirements engineering conference, 2009. rE'09. 17th iEEE international (pp. 24–33). IEEE.

Alspaugh, T. A., Scacchi, W., & Asuncion, H. U. (2010). Software licenses in context: The challenge of heterogeneously-licensed systems. Journal of the Association for Information Systems, 11(11), 730–755. Citeseer.

Bavota, G., Ciemniewska, A., Chulani, I., De Nigro, A., Di Penta, M., Galletti, D., Galoppini, R., et al. (2014). The MARKet for Open Source: An intelligent virtual open source marketplace. In IEEE cSMR-wCRE 2014 software evolution week (cSMR-wCRE).

Beardsley, M. C. (1950). Practical Logic. New York: Prentice Hall.

Bench-Capon, T., & Coenen, F. P. (1992). Isomorphism and Legal Knowledge Based Systems. Artificial Intelligence and Law, 1(1), 65–86.

Bench-Capon, T., & Gordon, T. F. (2009). Isomorphism and Argumentation. In C. D. Hafner (Ed.), Proceedings of the 12th international conference on artificial intelligence and law (iCAIL 2009) (pp. 11–20). New York, NY, USA: ACM Press.

Betz, G., & Cacean, S. (2011). The moral controversy about Climate Engineering - an argument map. Karlsruhe Institute of Technology. Retrieved from http://digbib.ubka.uni-karlsruhe.de/volltexte/1000022371

Chesnevar, C., McGinnis, J., Modgil, S., Rahwan, I., Reed, C., Simari, G., South, M., et al. (2006). Towards an argument interchange format. Knowledge Engineering Review, 21(4), 293–316. New York, NY, USA: Cambridge University Press.

Coleman, C. (2001, April). Lineo's gPL compliance tool. Retrieved from http://www.linuxdevcenter.com/pub/a/linux/2001/10/04/lineo.html

Dung, P. M. (1995). On the acceptability of arguments and its fundamental role in nonmonotonic reasoning, logic programming and n-person games. Artificial Intelligence, 77(2), 321–357. Essex, UK: Elsevier Science Publishers Ltd.

Fowler, M., & Scott, K. (2000). UML Distilled – A Brief Guide to the Standard Object Modeling Language (2nd ed.). Addison Wesley Longman, Inc.

Freeman, J. B. (1991). Dialectics and the Macrostructure of Arguments: A Theory of Argument Structure. Berlin / New York: Walter de Gruyter.

Germán, D. M., & Di Penta, M. (2012). A method for open source license compiliance of java applications. IEEE Software, 29(3), 58–63.

Germán, D. M., D. M.n, Di Penta, M., & Davies, J. (2010). Understanding and auditing the licenses of open source software distributions. In Proceedings of the 2010 iEEE 18th international conference on program comprehension (iCPC 2010) (pp. 84–93).

Gordon, T. F. (2011). Analyzing Open Source License Compatibility Issues with Carneades. In Proceedings of the thirteenth international conference on artificial intelligence and law (iCAIL-2011) (pp. 50–55). ACM Press.

Gordon, T. F., Governatori, G., & Rotolo, A. (2009). Rules and Norms: Requirements for Rule Interchange Languages in the Legal Domain. In G. Governatori, J. Hall, & A. Paschke (Eds.), Rule representation, interchange and reasoning on the web, LNCS (pp. 282–296). Berlin: Springer.

Gordon, T. F., Prakken, H., & Walton, D. (2007). The Carneades Model of Argument and Burden of Proof. Artificial Intelligence, 171(10-11), 875–896.

Hart, H. L. A. (1961). The Concept of Law. Oxford: Clarendon Press.

Hoffmann, M., & Borenstein, J. (2014). Understanding ill-structured engineering ethics problems through a collaborative learning and argument visualization approach. Science and Engineering Ethics, 20(1), 261–276. Springer Netherlands. Retrieved from http://dx.doi.org/10.1007/s11948-013-9430-y

Hohfeld, W. N. (1913). Some Fundamental Legal Conceptions as Applied in Judicial Reasoning. Yale law journal, 16–59.

Klein, M. (2011). How to harvest collective wisdom on complex problems: An introduction to the mit deliberatorium. Center for Collective Intelligence working paper.

Kunz, W., & Rittel, H. W. (1970). Issues as elements of information systems. Institut für Grundlagen der Planung, Universität Stuttgart.

Leffingwell, D. (2010). Agile software requirements: Lean requirements practices for teams, programs and the enterprise. Addison-Wesley.

LIDESC (2001). LIDESC: Libroch license awareness system. Retrieved from http://www.mibsoftware.com/librock/lidesc/

Loll, F., & Pinkwart, N. (2013). LASAD: Flexible representations for computer-based collaborative argumentation. International Journal of Human-Computer Studies, 71(1), 91–109. Elsevier.

OSLC (2007). The open source license checker. OW2 Consortium. Retrieved from https://wiki.ow2.org/oslcv3/

Prakken, H. (2010). An abstract framework for argumentation with structured arguments. Argument & Computation, 1, 93–124.

Tuunanen, T., Koskinen, J., & Kärkkäinen, T. (2009). Automated software license analysis. Automated Software Engineering, 16(3-4), 455–490. Springer.

Walton, D. (1996). Argumentation Schemes for Presumptive Reasoning. Erlbaum.

Walton, D., Reed, C., & Macagno, F. (2008). Argumentation Schemes. Cambridge University Press.

Chapter 11

Dialogues in US Supreme Court Oral Hearings

Latifa Al-Abdulkarim, Katie Atkinson, & Trevor Bench-Capon

Department of Computer Science, University of Liverpool, UK
{latifak, katie, tbc}@liverpool.ac.uk

Abstract. Dialogue protocols in Artificial Intelligence and Law have become increasingly stylized, intended to examine the logic of particular legal phenomena such as burden of proof, rather than the procedures within which these phenomena occur. While such work has provided some valuable insights, the original motivation still matters, and so in this paper we will return to the original idea of using dialogue moves to model particular procedures by examining some very particular dialogues - those found in oral hearings of the US Supreme Court. We will characterize these dialogues, and illustrate the paper with examples taken from a close analysis of a case often modelled in AI and Law, *California v Carney* (1985). This paper presents the preliminary investigation required to identify tools to provide computational support for the analysis of oral hearings.

1. Introduction

Dialogue games were originally introduced into AI and Law as a way of modelling legal procedures (Gordon, 1993), but more recently they have been used rather to capture the logic of aspects of legal reasoning, such as reasoning with cases (e.g., Prakken & Sartor, 1996) or particular legal phenomena such as burden of proof (e.g., Prakken & Sartor, 2007), and in consequence have become somewhat stylized and unrelated to any particular legal dialogue. That work has produced some valuable insights, but in this paper we will return to the original motivation and consider some particular dialogues that form a clearly defined stage of the US Supreme Court process, namely the Oral Hearings stage. Manual analysis of transcripts of these dialogues plays an important part in building systems to reason with cases in particular legal domains, such as CATO (Aleven, 1997) for U.S. Trades Secret Law. Our immediate aim is to present the preliminary investigation required to identify tools to provide computational support for the analysis of oral hearings: in the longer term we hope to provide a suite of tools to support other parts of the Supreme Court process. We begin by providing some necessary background. We will recall the notion of dialogue types used in (Walton & Krabbe, 1995), briefly describe the Supreme Court processes and the role played by the oral hearings, and describe the particular case we will use as a running example, *California v Carney* (471 U.S. 386 (1985)). The full transcript of the Oral Hearing and the opinions are available at http://holmes.oyez.org/cases/1980-1989/1984/1984 83 859.

1.1 Characterizing Types of Dialogue

When analyzing dialogues, it is important to be aware of the type of the dialogue, since shifts between dialogue types often lead to misunderstandings and fallacies; the dialogue types identify the speech acts available and provide the context to interpret them. Walton and Krabbe (1995) characterize dialogue types based on:

- The dialogue initial situation, which identifies the initial conditions that give rise to the dialogue.
- The overall collective goal, shared by all participants, which defines the characteristics of a successful dialogue outcome.
- The individual goals of the participants, which help to determine the reasons for particular move choices by the participants, which should lead towards the main goal, while at the same time respecting their own best interests.

In section 2.1 we will identify the initial situation and goals appropriate to Oral Hearings of the US Supreme Court, which will help to drive our analysis of the dialogues.

1.2 Supreme Court Process

Typically the Supreme Court reviews cases that have been decided in lower courts, either affirming or reversing the lower court decision. The Supreme Court receives a number of *certiorari* requests from parties who are not satisfied with lower court decisions asking for a review of their cases. Normally, when a case for consideration of *certiori* is accepted, the petitioner and respondent write briefs setting out their positions and recommendations to prepare the Justices for the oral argumentation. Briefs may also be supplied by other interested parties, such as the Solicitor General. These are the so-called *amicus curiae* (friend of the court) briefs. When the justices have considered all the briefs, the oral hearings take place. The total time for the oral argumentation is just one hour, thirty minutes for each party. Normally the petitioner will begin, reserving some of his thirty minutes for rebuttal. The respondent will follow for thirty minutes, and the petitioner will finish taking the remaining time for a rebuttal (cf the 3-ply structure of argumentation in HYPO (Ashley, 2009) and CATO (Aleven, 1997)). Following the oral hearing, the justices meet in a justice conference to discuss and vote on the case. Following this the opinions are prepared: one justice will be chosen to write the opinion of the Court, and the other justices may, if they wish, write their own concurring or dissenting opinions. Figure 1 below illustrates the procedure of the Supreme Court from the time of receiving the certiorari until the case is decided.

The Supreme Court is expected to give a decision in the case under review, but it needs to look to the past and the future as well. The decision needs to be expressed as a rule which will be applicable to future cases, and which will, as far as possible, be consistent with previous decisions of the Court: see e.g. Horty & Bench-Capon, 2012. The rule not only binds future courts, but provides guidance for those responsible for enforcing the law. Thus *Carney*, for example, provides police officers with a particular *test* to determine whether the automobile exception to the fourth amendment applies or not.

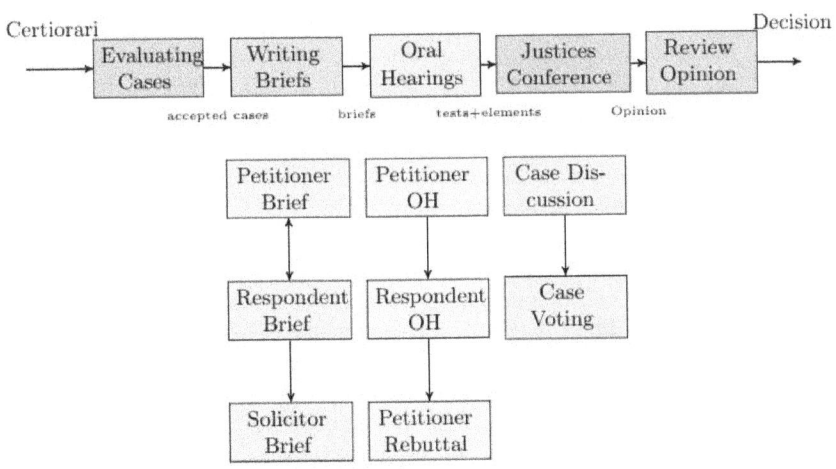

Figure 1. U.S. Supreme Court Procedure

1.3 A Case Study: California v Carney

This case is concerned with whether the exception for automobiles to the protection against unreasonable search provided by the Fourth Amendment applies to mobile homes, in particular motor homes in which the living area is an integral part of the vehicle. The Fourth Amendment protects the "right of the people to be secure in their persons, houses, papers, and effects, against unreasonable searches and seizures." A search is considered reasonable if a warrant has been obtained. The exception for automobiles was introduced in the *Carroll*[1] case in 1925, when a car carrying contraband was stopped and searched without a warrant. In *Carroll*, there is

---
[1] Carroll v. United States, 267 U.S. 132 (1925)

no mention of privacy: the justification was in terms of exigency, that the law could not be enforced if a warrant were required as the evidence would simply disappear into the night and another jurisdiction. The facts of this case (sedan on freeway) are as strong as can be in terms of exigency.

The notion of an automobile exception developed over the years through a series of cases, with elements of privacy being considered as well as exigency. *South Dakota v Opperman* (1976) gives a clear statement incorporating the reduced expectations of privacy appropriate to automobiles in addition to exigency, and this statement of the exception was used as the current rule by the majority in *Carney*. In *Opperman* an illegally parked car was impounded and searched without a warrant. Marijuana was found in the glove compartment. Note that in *Opperman* there was no exigency at all. *California v Carney* arose when drug agent officers arrested Carney who was distributing marijuana from inside a motor home parked in a public parking lot in the downtown of San Diego for unknown period of time. After entering the motor home, without first obtaining a warrant, the police officer observed marijuana. This motor home was an integral vehicle with wheels, engine, back portion and registered as a house car which requires a special driving license in California. On the other hand, it has some interior home attributes such as refrigerator, cupboard, table, scale, bag and curtains covering all the windows. The question was whether warrantless search was permissible in this case, satisfying the exception to the fourth amendment for automobiles. The California Superior Court *affirmed* the warrantless search because of the automobile exception. However, The California Supreme Court *reversed* the lower court decision indicating that the search violated the fourth amendment rule. After granting *a certiori*, The U.S. Supreme Court held that the search was reasonable and did not violate the fourth amendment rule and so reversed the California Supreme Court decision.

California v Carney has often been used in AI and Law to explore Supreme Court oral argument (e.g., Rissland, 1989; Walton & Krabbe, 1995), and to consider the interaction of two competing values (e.g., Bench-Capon & Sartor, 2003). In *Carney*, the competing values are enforceability of the law, which makes exigency important, and citizens' rights, which include the right to privacy (Bench-Capon, 2011).

2. Models of Reasoning with Cases in AI and Law

Modelling reasoning with legal cases has been a central topic of AI and Law from the beginning, and there is now a good degree of consensus, especially with regard to the main elements involved. This consensus can be expressed as a tree of inference with a legal decision as the root and with evidence as the leaves. Between the two we have a number of distinct layers.

Immediately below the decision we have a level of issues (Brüninghaus & Ashley, 2003), or values (Bench-Capon & Sartor, 2003), which provide the reasons why the decision is made. The idea here is that laws are made (and applied) so as to promote social values: whether a value is promoted or not is an issue. Where more than one value is involved and they point to different decisions, the conflict needs to be resolved. Sometimes it is appropriate to give priority to one value over another (as in Bench-Capon & Sartor, 2003), sometimes a balance needs to be struck (as in Brüninghaus & Ashley, 2003). Thus the Fourth Amendment exists to protect *Privacy*, and the automobile exception to enable *Law Enforcement*: in a particular case the issues will be whether there was sufficient exigency and/or insufficient expectations of privacy. Note that the relation between issues may be seen as a matter of ordering, or requiring a balance between the values: there is as yet no consensus on this point (Bench-Capon, 2011).

At the next level down there are a number of *factors* (Aleven, 1997). Factors are stereotypical fact patterns which, if present in a case, favor one side or the other by promoting a value, and so are used to resolve the issues. Factors are required to enable generalization across the infinitely varied fact situations that can arise, and so permit the comparison of cases. Sometimes (as in Aleven, 1997) it may be convenient to group several factors together under more abstract factors, so that we may have two or three layers of factors, moving from the base level factors through more abstract factors, before reaching the issues.

Below the factors we have the fact patterns used to determine their presence. These may offer necessary and sufficient conditions, but more often they offer either a set of sufficient conditions, or in less clear cut cases, a number of facts supplying reasons for and against the presence of the factor which need to be considered and weighed to make a judgment.

At the lowest level there is the evidence. Facts are determined by particular items of evidence, and where evidence conflicts a judgment will need to be made: often this judgment is made by a jury of lay people rather than lawyers. In the lower courts there will be real items of evidence, particular witness testimonies and the like. But by the time a case reaches the Supreme Court, the facts are usually considered established and beyond challenge. The Supreme Court does, however, need to consider what should count as evidence, and whether this will be generally be available, so that the rule can be applied in future cases. For example a birth certificate is normally required as evidence of age, other evidence being considered unreliable, or unlikely to be available.

Thus a complete argument for a case will comprise a view on what can be considered as evidence for relevant facts: what facts are required to establish the presence of various factors, and how they relate; how the factors can be used to determine the issues; and, where issues and values conflict, how these conflicts should be resolved. In the next section we will consider the individual and collective goals of the oral hearing dialogues.

2.1 Oral Hearings

In this section we will describe the initial situation, the individual goals and the collective goal for Oral Hearings in terms of the computational understanding developed in the previous section. As part of the Supreme Court procedure, there are three nested dialogues in the main oral argumentation dialogue. The overall aim is to establish the various elements, and the connections between them, expressed as clearly and unambiguously as possible, which can be used by the justices to construct the arguments they will use in their opinions. Each of the three dialogues will involve a counsel and nine justices. We will not distinguish between the justices here. Essentially they will all ask critical questions to clarify and challenge the argument elements proposed by counsel, although the particular questions they pose may well be motivated by their own views of the case, and their developing ideas of the argument they will use to decide the case. The arguments produced in the opinion will essentially use a *test*, which will be binding on future cases satisfying the test, and which will allow a decision to be made by using the facts of the current case, to establish the presence of a set of factors which will resolve the issues in favour of one of the parties.

In the *initial state* of the petitioner presentation, briefs from the petitioner, respondent and any "friends of the court" are available. These will set out (and justify) a set of tests which would provide candidate arguments: counsels will in turn present the elements of a test which, if accepted, will ground an argument for their clients. The briefs will also state the accepted facts of the case, and draw attention to relevant precedent cases. The *collective goal* is to obtain a clear statement of a set of elements that can form an argument which will resolve the case. Individually the *counsel* will wish to present an acceptable test which will lead to a decision for the petitioner and to answer any critical questions satisfactorily: modifying his tests if necessary. The *justices* will wish to clarify any points that had not been made clear in the original brief, and to pose challenges arising from the other briefs.

The *collective goal* of the second dialogue, the respondent presentation, is to obtain a clear statement of the test advocated for the respondent. The respondent dialogue differs in its *initial state* because the petitioner has already presented. Thus as well as presenting his own test, *counsel* for the *respondent* may wish to find some difficulties with the test proposed by his adversary. The *justices* remain interested in clarification and eliciting answers to questions arising from the other briefs.

While the *collective goal* of the rebuttal dialogue is again a clear statement of the tests and the elements composing them, and the choices that must be made when deciding between the tests, the *initial state* now also contains the respondent's test and its elements, and the individual goal of the petitioner's *counsel* is to pose questions against this test. *Justices* usually say very little during this stage, but they may wish to seek clarification of some points.

The goal of the three dialogues as a whole is to provide a clear statement of possible tests and the elements used in them which the justices will employ to decide the case and construct the arguments in their opinions.

2.2 Dialogue Moves for Oral Hearings

The goals of the dialogues involve identifying the elements that can be used to construct tests that will provide arguments to resolve the case, and the relationships between these elements. Speech acts will thus enable the proposal of these elements, and a set of critical questions challenging the elements, or seeking additional elements. Thus, although there is no conclusion, and hence no argument as such, the moves have many similarities to those arising from argumentation schemes. In this section we describe the moves, (each illustrated with an example from *California v Carney*). Formal definition of these moves will form part of the specification of a computational tool, which will be our next step.

- **Values Assertion:** The following values are relevant to decide the legal question. *Law Enforcement and Privacy are the values relevant to determining whether a case falls under the automobile exception.*
- **Issues Assertion:** The values are considered as these issues. *The issues are whether there was sufficient exigency (so that Law Enforcement is promoted) and insufficient expectations of privacy (so that privacy is not demoted) to permit a search without a warrant.*
- **Issues Linkage Assertion:** The issues should be considered collectively as follows. *The issues are related as Sufficient Exigency ∨ Insufficient Privacy.*

  We then have a number moves to introduce factors relating to the issues.
- **Factors for Issue Assertion:** The following factors are relevant to resolving the issue. *Vehicle Configuration and Location are relevant to resolving Sufficient Exigency.*
- **Factor Linkage Assertion:** The factors relevant to the issue should be considered collectively as follows. *Sufficient Exigency is resolved by considering Vehicle Configuration ∧ Location.*

  Finally we need a number assertions to identify the facts relevant to the various factors:
- **Facts for Factor Assertion:** The following facts are relevant to determining whether a factor is present. *Wheels and Means of Propulsion are relevant to determining Vehicle Configuration.*
- **Fact linkage Assertion:** The facts relevant to the issue should be considered collectively as follows. *The presence of Vehicle Configuration is determined by considering (Wheels ∧ Engine) ∨ (Boat ∧ (Engine ∨ Oars ∨ Sail)).*

Note that we do not need to consider the evidence level: the facts to be used have already been determined by the lower court, although, as we shall see from the CQ10 below, we do need to consider how the facts will be determined in practice.

We can now consider the critical questions that can be posed against these assertions. The structure as a whole is meant to provide *a test*. The questions relate

199

to the *test too broad* and *test too narrow* arguments of Ashley (2009), but our more articulated moves offer a finer granularity since they identify various different aspects with respect to which the test may be deficient. In the following CQs, by *relevant* we mean *relevant to deciding the case*.

CQ1: Are all the issues relevant?
CQ2: Are there other issues that are relevant?
CQ3: Are the issues linked correctly?
CQ4: Are all the factors really relevant to this issue?
CQ5: Is there an additional factor relevant to this issue?
CQ6: Is the relationship between factors correct?
CQ7: Are all the facts relevant to determining the presence of this factor?
CQ8: Is there an additional fact relevant to the presence of this factor?
CQ9: Is the relationship between facts correct?
CQ10: Can these facts be observed by the appropriate person?

These CQs permit a test to be challenged as too broad or too narrow at all three levels, and in two ways. As well as challenging the breadth and narrowness in terms of the elements used (e.g. CQ1 and CQ2), the breadth and narrowness can also be challenged in terms of the way the elements are combined, as in CQ3. It should also be noted that it is quite common to combine questions: for example CQ1 and CQ2 can be combined, effectively suggesting the substitution of one element for another. These could be expressed as additional CQs, but here we will rely on combinations of CQs. Note also that CQ10 relates to whether the tests can be applied by the person responsible for applying them in the operational situation: a test that cannot be applied in the actual situation is not acceptable, because ensuring that the test will be applicable in future cases is essential.

In Ashley (1990) the response to such questions is said to be one of:

> **Save the test:** Effectively deny that the question is pertinent to the test; for example if CQ8 is posed suggesting that an additional fact would change the position with respect to some factor, it can be maintained that the same position continues to hold.
> **Modify the test:** Exclude an item (e.g. CQ1), add an item (e.g. CQ2) or change the linkage (e.g. CQ3);
> **Abandon the test:** This means withdrawing the current proposal and proffering a new one.

In the course of the hearing the various elements of the proposed tests emerge. The dialogue is often not well structured: the questions are not posed in any particular order, and may be interleaved with the presentation of the proposal, so that the proposal is modified as it is presented. None the less, the aim of each counsel is to present and defend the elements required for a test which will decide the case for their client, and the justices aim to get a clear statement of the various elements which they can use to build the arguments in their opinions.

3. Illustrations with *California v Carney*

Using the set of moves proposed above, we have analyzed the oral hearings of *California v. Carney*. There is insufficient space to report the full analysis here, but we provide example extracts from each of three component dialogues, showing the moves proposed and some critical questions posed against them.

3.1 Dialogue One - Petitioner Oral Hearing

The petitioner maintains throughout the dialogue the position that exigency is the only issue here, and that the relevant abstract factor of *inherent mobility*, that is, the capability of quickly becoming mobile, using the configuration of the vehicle as the base factor, ensures a sufficient exigency for the automobile exception to override the privacy protection of the fourth amendment. He proposes that this can be determined using easily observed facts such as the vehicle *has wheels* and the *vehicle is self-propelled*. One justice poses CQ7, suggesting that wheels might be enough, so that trailers are also covered, but the counsel rejects this suggestion and maintains his test. The question of boats is also raised (CQ8) suggesting that there must be some other consideration to cover boats. In this case the test is modified to disjoin boat with oars to provide an additional sufficient condition (CQ9).

Figure 2 illustrates some of the elements that make up the petitioner's test. An important feature of the argument is that the issues of privacy and exigency are kept separate. A justice challenges this, using CQ3, based on the Solicitor General's brief, which suggests that both must be considered.

> **Unidentified Justice:** You prefer a single rationale for the exception to the warrant requirement. Namely, you think "mobility" is practically the sole criteria; and the Solicitor General at least thinks that there are two.
> **Petitioner:** Well, I think there is more than one, and I think they're independent of one another,

The Solicitor General argued that there are some circumstances where a mobile home results in expectation of privacy (privacy issue) that must be considered in addition to exigency (CQ3). One example of these circumstances is when the motor home is stationary in a mobile home park for a significant period of time (CQ5). The petitioner rejects the use of the length of time parked as this cannot be determined by the law enforcement officer (CQ10). The notion of location is, however, accepted as a factor additional to vehicle configuration relating to exigency (modifying the test in response to CQ5), claiming that while a vehicle in a residential location (such as a mobile home park) *might* not be considered inherently mobile, whereupon issues of privacy would become relevant, a vehicle in a regular parking lot can always be considered inherently mobile.

201

Unidentified Justice: Well, anyway, you certainly would differ with the Solicitor General as to the application of the exception in a park, in a mobile home park?

Petitioner: Under the circumstances that's been presented, yes, I would.

Unidentified Justice: Of course that isn't the issue here, is it? This is in a public parking lot.

Petitioner: That's correct, Your Honor. That is not presented in this case.

And if I might address the Solicitor General's position and explain why ours is a little bit different: The reason for our difference with the Solicitor General is because that a law enforcement officer in the field has to determine whether or not this vehicle is now placed in a constitutionally protected parking spot: if an individual is going to come upon this vehicle he's not going to know whether it's been parked in this particular motor home lot for a period of three months, or two weeks, or how long.

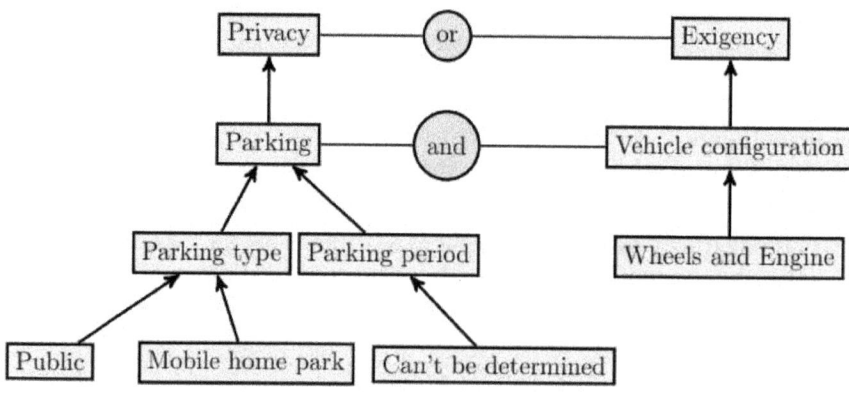

Figure 2. Elements Used in Petitioner's Test

3.2 Dialogue Two - Respondent Oral Hearing

The respondent presents a rather different test. He insists that both *exigency* and *privacy* need to be considered. He aims to establish sufficient expectations of privacy on the grounds that the motor home is designed to be used for residential purposes as well as transportation unlike a regular passenger automobile *(However, transportation is not its sole function)* and this can be determined by the presence of facts about configuration such as: *cab and the living quarters are part of a single unit.* Also to be considered are whether there are attributes associated with a home, established using facts such as *containing a bed and a refrigerator* and whether it is used to store and transport personal items[2] Additionally he aims to show that the exigency is lessened since the vehicle is not ready to be moved *(is inoperable)*, because there is no driver in position, and there are curtains drawn over the windscreen. Moreover because it was parked in downtown San Diego a warrant could easily have been obtained. The respondent's test is viewed as a rebuttal of the petitioner's argument, adding privacy as an issue (CQ2), and conjoins rather than disjoins the two issues (CQ3). *Readiness to move and possibility of obtaining a warrant* are introduced as additional factors for assessing exigency (CQ5). The additional *factors capable of use as a home* and *used to store and transport personal items* are proposed as additional factors relating to privacy (CQ5). The relationship between these factors was not discussed, although CQ6 was available to clarify it had any justice wished to do so.

The extract below is an attempt by a justice to suggest that the fact that the cab and the residential part are a single unit is not essential (CQ7) for the required vehicle configuration factor to be present. The respondent saves his test by pointing out that if we have a trailer and tractor configuration, the greater expectations of privacy apply only to the trailer, and the exigency applies only to the tractor. Under pressure, however, the respondent eventually modifies the test by introducing the personal effects factor (CQ5). This extract is summarized graphically in figure 3.

> **Unidentified Justice:** Assume now that the automobile vehicle is the tractor that would pull the otherwise immobile motor home, or whatever you want to call it. Now you could search the tractor, but not the–
> **Respondent:** I think that's true. And the reason is–
> **Unidentified Justice:** –The tractor can take off down the street and go 70 miles an hour on the highway?
> **Respondent:** –The reason is, the tractor has a privacy interest which society is less prepared to recognize. It's a diminished privacy expectation, as opposed to the motor home or the trailer itself.
> **Unidentified Justice:** Well, they're equally... when they're attached, they're equally moveable, aren't they?
> **Respondent:** Exactly. But one is used for private living residential purposes, and the other is used for transportation. As a matter of fact–
> **Unidentified Justice:** The other one isn't used for transportation in the abstract, but only in connection with what it pulls. Isn't that so?
> **Respondent:** –Yes, that's correct. Unidentified Justice: People don't go out on the highway on the tractor alone, do they?
> **Respondent:** Ordinarily not. The tractor partakes more of the automobile, because it doesn't have... it is not the kind of repository for personal effects.

---

[2] This is intended to align it with precedent cases involving luggage being transported in a car where the automobile exception did not apply. For example in U.S. v. Chadwick 433 U.S. 1 (1977) it was held that the warrantless search of a footlocker in the trunk of a car was reasonable.

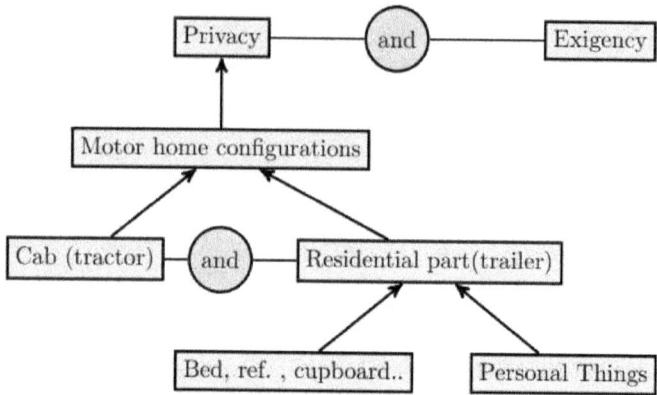

Figure 3. Some of The Elements Used in Respondent's Test

3.3 Dialogue Three - Petitioner Rebuttal

In the last of the three nested dialogues, the petitioner attempts to rebut the tests introduced by the respondent by adding new facts and/or factors or showing the inapplicability of the tests to prove sufficient privacy.

According to the respondent test above, that the living quarters are an integral part a vehicle should attract sufficient privacy expectations. However, in the following extract the petitioner claims that it is not possible to determine these residential facts (CQ10), and in this particular case there was no evidence of food or personal items inside the motor home. So even if the personal effects factor were relevant, that is (the respondent's CQ5) succeeds, it does not apply to *Carney*.

> But the record does not at all support this particular assertion.
> And in particular, if one examines the photographs that are a part of the record in this case that were submitted to this Court, looking at the picture of the refrigerator will show that there is marijuana in the refrigerator, but there is no food.
> And when they examined the cupboards in this case, there's no underwear, there's no sheets, there's marijuana.
> There's nothing in the record to suggest Mr. Carney was using this as his home, and in fact that is the problem.
> There is no way to determine, in these particular classes of vehicles, when they are and are not being utilized as a home, objectively.

4. Concluding Remarks

In this paper we have considered an important class of legal dialogues relating to reasoning about cases, namely the US Supreme Court Oral hearings.
We have:
- Located these dialogues in the overall Supreme Court process;
- Identified that the dialogue consists of three distinct sub-dialogues;
- Characterized the three sub-dialogues in terms of their initial state, and individual and collective goals;
- Presented a set of moves designed to enable the goals of the dialogues to be achieved in the form of assertions and associated critical questions;
- Illustrated all points throughout using extracts from the transcript of a case much discussed in the AI and Law literature.

In future work we will present our full analysis of the transcript of *Carney*; and apply the analysis to other related cases (e.g. those discussed in Bench-Capon, 2011). We will then specify a tool that will support the analysis of Oral Hearings by automatically constructing the corresponding tree from a transcript annotated with our moves. We will next relate the argument components that emerge from the Oral Hearing to the arguments that are expressed in the opinions of the Justices. We conjecture that the move from oral hearing to opinion will involve selection between the options and justification of the choices made. This last point may be of relevance for work on argumentation schemes since it provides a reasoned acceptance or rejection of critical questions, a topic which is as yet relatively unexplored in computational argument.

References

Aleven, V. (1997) Teaching case-based argumentation through a model and examples. PhD thesis, University of Pittsburgh.
Ashley, K. D. (1991). *Modeling legal arguments: Reasoning with cases and hypotheticals*. MIT press.
Ashley, K. D. (2009). Teaching a process model of legal argument with hypotheticals. *Artificial Intelligence and Law, 17*(4), 321-370.
Bench-Capon, T., & Sartor, G. (2003). A model of legal reasoning with cases incorporating theories and values. *Artificial Intelligence, 150*(1), 97-143.
Bench-Capon, T. J. (2011, December). Relating Values in a Series of Supreme Court Decisions. In *JURIX* (pp. 13-22).
Bench-Capon, T., & Prakken, H. (2010). Using argument schemes for hypothetical reasoning in law. *Artificial Intelligence and Law, 18*(2), 153-174.

Brüninghaus, S., & Ashley, K. D. (2003, June). Predicting outcomes of case based legal arguments. In *Proceedings of the 9th international conference on Artificial intelligence and law* (pp. 233-242). ACM.

Gordon, T. F. (1993, August). The pleadings game: formalizing procedural justice. In *Proceedings of the 4th international conference on Artificial intelligence and law* (pp. 10-19). ACM.

Horty, J. F., & Bench-Capon, T. J. (2012). A factor-based definition of precedential constraint. *Artificial Intelligence and Law*, *20*(2), 181-214.

Prakken, H., & Sartor, G. (1996). A dialectical model of assessing conflicting arguments in legal reasoning. In *Logical Models of Legal Argumentation* (pp. 175-211). Springer Netherlands.

Prakken, H., & Sartor, G. (2007, June). Formalising arguments about the burden of persuasion. In *Proceedings of the 11th international conference on Artificial intelligence and law* (pp. 97-106). ACM.

Rissland, E. L. (1989, May). Dimension-based analysis of hypotheticals from supreme court oral argument. In *Proceedings of the 2nd international conference on Artificial intelligence and law* (pp. 111-120). ACM.

Walton, D., & Krabbe, E. C. (1995). *Commitment in dialogue: Basic concepts of interpersonal reasoning*. SUNY press.

Chapter 12

An Experience in Documenting Medical Discussions through Natural Argumentation

Daniela Fogli[1], Massimiliano Giacomin[2], Fabio Stocco,[3] & Federica Vivenzi[4]

Department of Information Engineering, University of Brescia, Brescia, Italy;
[1]daniela.fogli@unibs.it
[2]massimiliano.giacomin@unibs.it
[3]fabio.stocco85@gmail.com
[4]federica.vivenzi@gmail.com

Abstract. The paper presents a prototypical tool to support the documentation of medical discussions. The development of this tool came after a long phase of requirement specification and prototyping activity, which allowed identifying two main functionalities for the tool, corresponding to two components. A first component is devoted to support the user in documenting clinical discussions, exploiting a graphical notation tailored to physicians' habits. A second component exploits argumentation schemes in order to analyze documented discussions, possibly bringing to light weaknesses in the reasoning process.

1. Introduction

Medical activity often requires different specialists to participate in meetings, in order to discuss cases, diagnoses or methodologies to adopt. This may happen periodically, when the hospital structure hosts patients that must be assisted over long periods of time, as in the case we analyzed during our collaboration with the Department for Disabled People of an Italian hospital. Here, patients affected by a variety of physical and cognitive disabilities are hosted in a set of home units, where they are assisted in their daily activities and, in some cases, are supported through a rehabilitation process; patient disease or healing evolution are regularly discussed in physician meetings. But meetings among different medical specialists also take place in every hospital ward in case of difficult diagnoses, interesting cases or rare pathologies.

However, as witnessed by many physicians we have interviewed, these discussions are never documented, neither on paper nor in electronic documents: medical specialists analyze symptoms, propose hypotheses, read examination results and, through a progressive development of the discussion, arrive for example at suggesting to adopt some drug treatment or to acquire more data through a further examination; only the decisions taken at the end of the discussion are traced in paper or electronic patient records.

In this paper, we describe a prototypical tool we have designed in collaboration with representative users, which aims at supporting the documentation of medical discussions. The purpose of this tool is two-fold: 1) it should allow physicians to recall a previous discussion, having a structured view of the alternative hypotheses taken into consideration and of the reasons why they were accepted or discarded, in order to continue the discussion in the light of new information concerning for example new examination results or the evolution of a patient's clinical condition; 2) it should help physicians identify weaknesses in the reasoning process, e.g. discovering that an alternative diagnosis has not been taken into consideration, or that an available treatment has been neglected, or that a decision taken should be reconsidered in the light of a particular patient's condition.

Our goal is to provide a tool that allows documenting argumentation processes, without forcing users to know the abstract theory necessary to describe them. The idea is remaining as much as possible close to the considered domain, namely to its vocabulary and best practices, and thus making argumentation a natural process.

The paper is organized as follows. Section 2 outlines the proposed approach and compares it with related work in the field of Clinical Decision Support Systems and computer-supported argumentation visualization. Section 3 describes the component of the tool devoted to the discussion documentation, while Section 4 presents the component devoted to discussion analysis. Finally, some brief conclusions are drawn in Section 5.

2. Outline of the Proposed Approach and Related Work

Problems that teams of physicians often have to face can be regarded as ill-structured (or wicked) problems (Rittel & Webber, 1973). Among others, several characteristics of wicked problems are worth mentioning for the medical domain. Problem specification is often ambiguous and incomplete, every problem is essentially unique and there is not a predefined path that could be followed from the problem to the solution, but rather problem solution may require several iterations, according to a trial and error approach. Moreover, multiple information sources are necessary to acquire all the knowledge related to the problem; particularly, in the medical domain, knowledge is very huge, in continuous evolution, incomplete, uncertain, inconsistent, vague, and heterogeneous (Cortés, Vàzquez-Salceda, Lòpez-Navidad, & Caballero, 2004). The search for a problem solution involves different stakeholders, with different culture and background; in the medical domain, collaboration among different specialists, with different competencies in a variety of fields, is often essential to arrive at a correct diagnosis.

Wicked problems are often dealt with the help of Decision Support Systems (DSSs) (Burstein & Widmeyer, 2007; French & Turoff, 2007), which are interactive software systems that provide decision makers with all the information related to the case at hand, including suggestions for actions in response to events, and make it possible to explore available data from different points of view and through different visualization tools, as well as to simulate scenarios that may occur as a consequence of some decision. In the medical domain, a variety of Clinical Decision Support Systems (CDSSs) have been proposed. CDSSs are very complex systems that must facilitate the coordination of the activities of different specialists and help them manage a huge amount of information, coming from heterogeneous sources, such as clinical trials, guidelines, historical data, and best practices (Wyatt & Sullivan, 2005; Fogli & Guida, 2013). CDSSs include systems for the management of electronic medical records, such as WebPCR (Web Browser based Patient Care Report), LifeLines (Plaisant et al., 1998), and CareVis (Aigner & Miksch, 2006), which allow monitoring the state of patients under specific medical treatments. There are also systems that, beyond presenting information, provide also suggestions about the decisions to take (Peleg, Wang, Fodor, Keren, & Karnieli, 2006); whilst others, such as REACT (Glasspool et al., 2007) or HT-DSS (Fogli & Guida, 2012), aim at helping physicians perform complex planning activities. Finally, there are also some CDSSs, such as CAPSULE (Walton et al., 1997), which provide justifications of suggestions.

However, the above CDSSs do not encompass adequate tools to track collaborative decision making processes; they provide classic communication methods, such as email or chat, which do not allow structuring the discussion, and which require that all exchanged messages are examined from the beginning, whenever decisions and motivations underlying them are to be recalled after a certain period of time.

In order to develop a tool to document clinical discussions, we have carried out a long phase of requirement specification and prototyping activity. The starting point was the analysis of a video taken from a meeting of a group of specialists discussing the case of an old patient affected by disorientation problems. The repeated examination of the video allowed identifying the discussion objectives, the kind of information that are exchanged, the structure and the elements of a discussion, as well as how a discussion is managed. Participants in the discussion usually propose some hypothesis, for example "Mr. T. might be affected by a cognitive degeneration", by sustaining it with information coming from direct observation (e.g., "Mr. T. does not find his room anymore"), from examinations, or from other information sources. In other words, some physicians provide their arguments to sustain a hypothesis, whilst the others usually propose counterarguments to attack that hypothesis, by basing their reasoning on further information sources or on a different interpretation of existing data (e.g., "Mr. T. could have a vision problem"). This "verbal and social activity of reasoning aimed at increasing (or decreasing) the acceptability of a controversial standpoint" (van Eemeren et al., 1996) is what is usually known as *argumentation*.

Existing systems implementing argumentation mechanisms have been thus analyzed, in order to derive useful hints for the design of our tool for medical discussion documentation. Most studies concerning these systems are mainly focused on dialogue protocols and on algorithms ruling the behavior of software agents, rather than on the representation of the process human users carry out in collaborative decision making; indeed, in this case, a suitable representation of arguments must be studied, and language and interaction style of the system must be tailored to users' needs, preferences, and cultural background. To this purpose, we have focused on the field of computer-supported argumentation visualization. According to (Karamanou, Loutas, & Tarabanis, 2011) and (Buckingham Schum, 2003), computer-supported argumentation visualization systems must present arguments in a clear and simple way, by adopting a language close to natural language and offering users an intuitive and easy-to-use interface. These systems are usually designed for a specific application domain (e.g., education, law, politics) and provide different kinds of diagrammatic representations of arguments, based on graphic notations proposed in argumentation theory. An interesting overview of such systems is presented in (Karamanou et al., 2011); among these systems, we have analyzed ArgVIS (Karamanou et al., 2011), Araucaria (Reed & Rowe, 2004), Rationale (Sbarski, van Gelder, Marriott, Prager, & Bulka, 2008), SEAS (Lowrance et al., 2008), Compendium (Compendium web site), and Carneades (Gordon, 2010).

In the context of the present paper, these systems can be classified along two dimensions, i.e. the purpose of argument visualization and the graphical language exploited for this task.

As far as the first dimension is concerned, argument visualization can be applied to two purposes at least. On the one hand, some tools focus on providing a structured representation of the arguments found in a textual document, with the aim of identifying and analyzing them. This is the case of Araucaria (Reed & Rowe, 2004), whereby a user is able to select portions of text from a written document that can then be referred to in a diagrammatic representation of the arguments. On the other hand, some tools are more oriented on driving a discussion, allowing the participants to have a clear view of the arguments advanced so far and supporting them in the debate. For instance, both Compendium (Compendium web site) and Carneades (Gordon, 2010) provide the user with a set of argumentation schemes that can be exploited as patterns for the construction of new arguments as well as for the identification of the relevant counterarguments (Reed and Walton, 2003; Walton, 2006). A similar view is enforced in ArgVIS (Karamanou et al., 2011), which also allows the users to interactively modify a debate graph with different privileges. The distinction pointed out above is called "argument as product" vs "argument as process" in (Reed and Walton, 2003).

As for the second dimension, a variety of diagrammatic languages are exploited to represent the arguments. For instance, Araucaria (Reed & Rowe, 2004) supports Wigmore diagrams (Wigmore, 1913), Toulmin's notation (Toulmin, 2003) as well as a standard notation, while the languages adopted in Rationale (Sbarski et. al.,

2008), ArgVIS (Karamanou et. al., 2011) and Compendium (Compendium web site) are IBIS-like (Werner & Rittel, 1970).

Since the early phases of the iterative development of the tool, some user requirements have been identified with respect to the characteristics above:

- Physicians are not willing to follow any discussion protocol, but they want to feel free to participate in the discussion according to their usual habits. For instance, sometimes they want to point out all of a patient's symptoms, other times they want to focus on a subset of them to identify a diagnosis, other times they tentatively reason about a diagnosis and look for the corresponding symptoms.
- Physicians adopt a specific medical terminology with a shared meaning, and do not accept to characterize propositions in abstract ways, e.g. identifying a major premise w.r.t. a minor one, or distinguishing between data and general rules.
- Even though physicians interact by pointing out arguments and counterarguments, they are not willing to make the relevant structure explicit during the discussion, let alone conform to a predefined scheme.
- Physicians require a structured representation of a previous discussion to somewhat adhere to the way discussion has been carried out. In particular, they do not accept to structure the information according to a predefined scheme if this does not reflect the order in which information has been pointed out. For instance, if a hypothetical diagnosis has been proposed before looking for symptoms, they do not accept a discussion representation where this order is reversed, e.g. presenting the symptoms first and then a diagnosis as a possible cause.
- Physicians require some information to be grouped according to specific needs, e.g. all clinical examinations and related results should be visualized together.
- The language used to document the clinical discussion must be clear, easy to understand and specific to the medical domain.

While the above requirements leave little space for *documenting* a clinical discussion by means of the graphic languages proposed in argumentation theory, we have experimented their adoption to *analyze* the discussion "a posteriori". In particular, we have exposed physicians with a prototypical medical discussion represented in several notations and got their feedback. Wigmore diagrams (Wigmore, 1913) turn out to be difficult to understand, while Toulmin's model (Toulmin, 2003) appears somewhat abstract, in particular evidencing the distinction between its components is considered unnecessary. On the other hand, physicians have recognized the usefulness of argumentation schemes (Walton, 2006) specifically devised for the clinical domain, similar to those exploited in Carrell+

(Tolchinsky, Cortés, Modgil, Caballero, & Lòpez-Navidad, 2006). Argumentation schemes turn out to be easily understandable, and physicians agree that they may be a valid instrument to identify weaknesses in the discussion, e.g. that a possible diagnosis has been neglected, that a doubt about an accepted diagnosis may be raised, that a clinical test should be prescribed, or that a possible treatment has not been taken into account.

On the basis of these considerations, the tool has been structured in two components.

A first component allows a trained user to document a clinical discussion by producing a graphical representation, possibly after its conclusion, on the basis of a video recording of the discussion or of paper-based notes. This choice is motivated by the fact that physicians often interact under critical time constraints, thus the use of the tool during the discussion may be regarded as time-consuming. In the iterative development of the first component, attention has been focused on the graphical language adopted to represent the discussion: on the one hand it must be tailored to the physicians' habits in order to fulfill the requirements presented above; on the other hand, it must guarantee a structured representation of the discussion in order to allow physicians to quickly recall it after some time.

A second component is devoted to the analysis of a discussion previously documented by means of the first component. This second phase is optional, and is delegated to an expert user (possibly the same as the user of the first component) familiar with argumentation schemes, which is allowed to select argumentation schemes from a repository and to instantiate them with the elements of the discussion (this somewhat resembles the use of Araucaria).

3. Creating Arguments in Medical Discussions

3.1. Documenting Medical Discussions

The part of the tool to be used for tracking and managing medical discussions has been developed through an iterative approach, including the design of paper-based and interactive prototypes and various interviews with representative users (students in medicine and physicians). This activity has led to define the terminology to be used in the system and to understand how to support the creation and modification of a discussion.

The idea is structuring each discussion about a clinical case as a tree diagram, somewhat resembling the IBIS-like notation of Rationale (Sbarski et. al., 2008), but adopting a specific medical ontology. More specifically, the tree diagram will include different kinds of nodes corresponding to the different medical concepts that physicians use during discussions (diagnosis, symptom, examination result, and so on). Therefore, users are not forced to use terms not familiar to them, such as

An experience in documenting medical discussions through natural argumentation

"argument", "counterargument", "support", "attack", even though they will implicitly express such kinds of concepts and relations during tree construction.

Figure 1 shows a screenshot of the resulting system. This first version of the system is in Italian to increase its acceptance by our users, but it will be described in the following by using English terms. The top bar includes the buttons for creating a new discussion, opening a previous discussion, saving the current discussion, and analyzing the discussion by activating the other component of the system. The main area is composed of three parts:

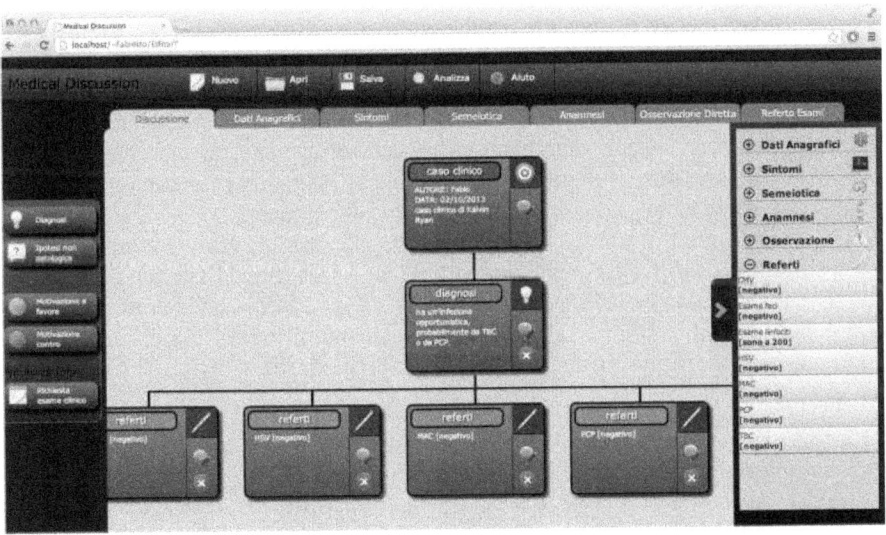

Figure 1. The part of the tool devoted to the documentation of medical discussions

1. A *left area*, which includes the 5 types of nodes – diagnosis, non-pathological hypothesis, motivation pro, motivation against, and examination request – that represent the basic element types of a discussion. The distinction among these element types is subtle and different physicians can classify an information item in different ways. On the other hand, the 5 types of nodes arose during requirement analysis as those physicians want to be available in the system. Each of the five nodes can be included in a discussion by dragging-and-dropping it in the central area (see below); node instantiation is carried out by the system by asking the user any useful content.
2. A *central area*, which is in turn divided in 7 tabs: Discussion, Personal Data, Symptoms, Semiotics, Case History, Direct Observation, and Examination Report. The "Discussion" tab will host the tree representing the discussion: the root node (Clinical Case) is generated automatically by the system when the user creates a new discussion; whilst, the other nodes are created by dragging and dropping the elements available in the left

and right areas. It has to be remarked that the links between nodes in the discussion tree do not have a precise meaning; they simply reflect the fact that a consideration in a discussion is referred to a previous one in some way. Therefore, users can never build an inconsistent or syntactically incorrect diagram. Information items can be organized according to well-founded structures in the analysis component (see Section 4). The remaining tabs correspond to the six different sources of information that can be used to support or attack the elements of the discussion. Indeed, information are usually gathered from the interviews with patient and patient's relatives (personal data, symptoms, and case history), from the detection of clinical signs by the physician during a physical examination (semiotics), from observations of patient behavior by physicians, nurses, social assistants, and other stakeholders (direct observation), and from results of clinical examinations (examination report). The tab order reflects the order followed by the physician to gather information before proposing a diagnosis or a non-pathological hypothesis.

3. A *right area*, which summarizes all the information gathered during the meeting, and whose details can be found in the tabs in the central area: each item can be selected and included in the "Discussion" tab to become an element in favor or against another element of the discussion, which is usually a diagnosis or a non-pathological hypothesis.

In Figure 1, the "Discussion" tab shows a discussion tree under creation, referring to a case from the well-known "House (M. D.)" TV series. House's team is talking about the problems of Kalvin Ryan, a young patient affected by HIV. Dr. House has suggested that the patient has an opportunistic infection, thus a diagnosis has been added under the tree root. Dr. Coleman has objected that the results of clinical examinations exclude this diagnosis; therefore, four nodes against the infection diagnosis have been dragged and dropped under the diagnosis node. These nodes have been created by selecting elements from the results of examination reports summarized in the right area.

3.1. *Evaluation with Users*

An expert physician, a novice physician and a senior student in medicine have participated in testing the current version of the prototype. After a brief training that illustrated how the system can be used to document the House's discussion described in the previous section, users have been required to document the subsequent part of the same discussion. A thinking-aloud observation method has been adopted to collect as much as possible information from users, namely users' reasoning strategies, terminology, and reactions to the system appearance and behavior.

The first part of the test consisted in a series of specific tasks extracted from the House's case, which allowed users to familiarize with the system and evaluators to identify its main usability problems. In the second part of the test, users have been required to autonomously create a discussion model; this has been useful to investigate if ambiguities and imprecisions affect the system from the point of view of the logical development of a discussion. In particular, we have observed that the two more experienced users (expert and novice physicians) had much less difficulties in performing the task with respect to the student. They moved easily among tabs and lateral areas for creating the various nodes, even though they tended to work out again the discussion content according to their experience and background.

It is interesting to note that when symptoms are described in the discussion, they are correctly added in the corresponding tab and then used as elements to substantiate some diagnosis or non-pathological hypothesis; whilst, when the absence of some symptoms is specifically mentioned in the discussion, this information is used in "motivation against" nodes. Even though this is perfectly coherent with the discussion logic, the expert physician raised some perplexities about this asymmetrical behavior; therefore, we plan to examine this situation in the future, possibly proposing a re-classification of concepts.

Notwithstanding the physicians operated freely on the discussion, the resulting diagrams presented some interesting regularities. For example, nodes at level 1 refer always to diagnoses or non-pathological hypotheses, and actual chains of reasoning are created after these nodes. The tree representation shows its effectiveness when a diagnosis is progressively refined through more specific diagnoses, as a consequence of the consideration of additional information. For instance, Figure 2 presents an excerpt of the discussion model created by the novice physician, where three diagnosis nodes are nested from top-left to bottom-right and a forth node in the chain includes the request of a further examination whose results could substantiate the last and most probable diagnosis.

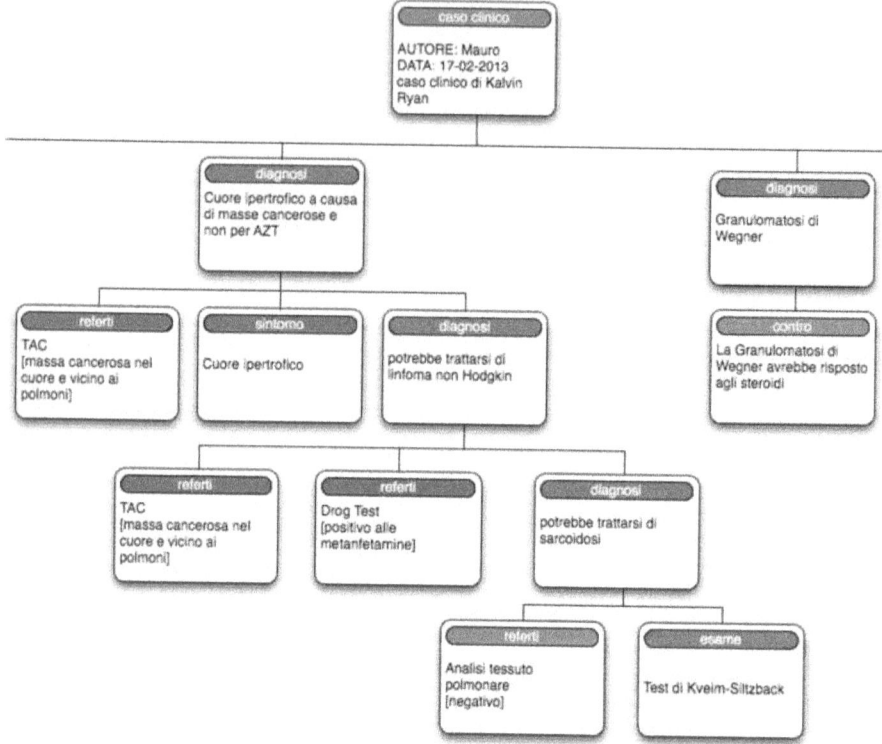

Figure 2. Excerpt of a medical discussion described by a physician using the tool

In a final brief interview, users have confirmed that the tool could be very useful for documenting medical discussions. According to the users, the system may facilitate the comprehension of a discussion and of the decisions taken, thanks to the schematic and structured model that the system allows one to create. Users also added that the system could help physicians arrive at more robust diagnoses in less time. Finally, users said they found the system easy to use and to learn, by clarifying that the graphical representation of the discussion stimulates the participation in the modeling activity.

4. Analysis of Discussions Based on Argumentation Schemes

The component devoted to the analysis of a discussion exploits a different and more synthetic visualization of the discussion, based on a multi-level list and no more on a tree diagram. In this way, more space is reserved to the content of the discussion and navigation can be limited to a vertical scroll. Figure 3 shows a screenshot of this part of the system. The main area is composed of three tabs, called Discussion,

Arguments, and Excluded Items. When the user accesses the first tab s/he can walk through all the elements of the discussion and access their details. Here the user can add an argument by selecting the button "New Argument" on top of the screen. This selection activates a popup window that asks the user the argument title and to choose, from an available list, the argumentation scheme s/he would like to use for the argument under creation. After this selection, the user can identify with a simple click on the list items those elements that can be regarded as the argument premises or conclusion. In particular, on the basis of the analysis of the recorded video and inspired by the work of Walton (Walton, 2006), we have defined five argumentation schemes suitable to the medical domain. However, these schemes represent only a preliminary proposal, which deserves further investigations in the future. All the arguments created by the user with reference to a discussion are summarized in the right area of the screen; here, arguments are classified on the basis of the argumentation schemes of which they are instances. Furthermore, each argument selection in the right area makes related premises and conclusion appear in the multi-level list.

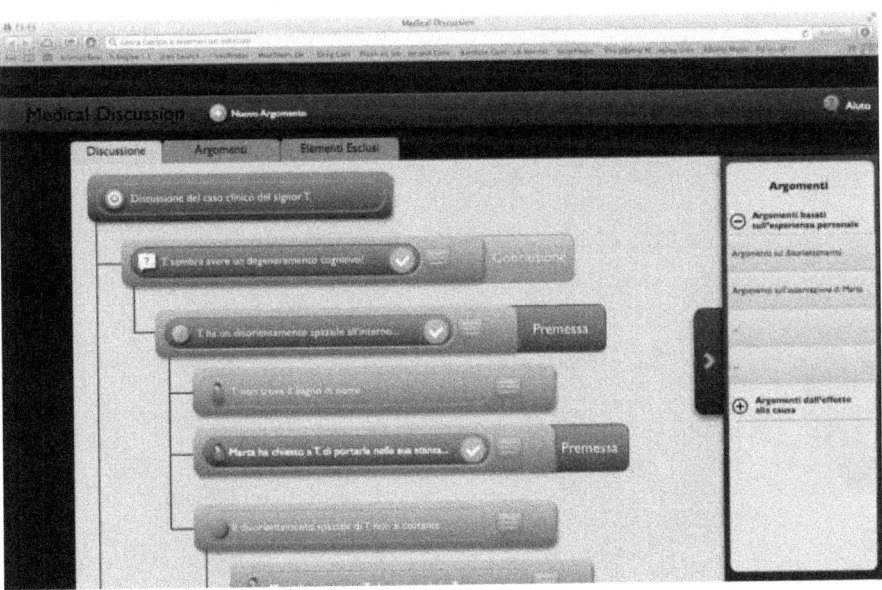

Figure 3. The part of the tool devoted to the analysis of medical discussions

The "Arguments" tab (Figure 4) shows the details of the arguments identified in the discussion: the left part presents the list of the arguments, classified again according to the argumentation schemes; the right part provides the details of a selected argument, namely its Premises, Conclusion, and Critical Questions.

Let us note that the critical questions are those associated to the argumentation scheme chosen by the user during argument creation; critical questions are properly instantiated by using data in argument premises and conclusion. For each critical question, the user can indicate if an answer to the question already exists (green checkmark), or if the answer is still missing (red cross sign).

The last tab ("Excluded Items") gathers all discussion elements that do not belong to any arguments. In this way, the user is provided with further information that s/he might consider for deciding about the answer to critical questions.

The part of the system devoted to the analysis of discussions has not been tested with users yet. We hypothesize that a team member, trained in basic notions of argumentation theory and argumentation schemes, could use it after a discussion has been documented. In this way, this expert could help the other members of the team identify weaknesses in the discussion, and suggest further investigations where answers to critical questions are missing.

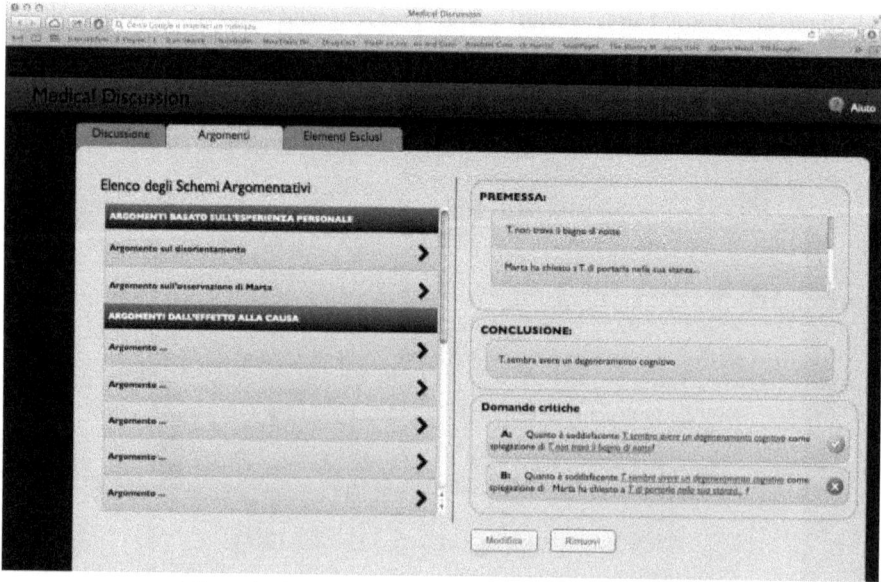

Figure 4. The tab of the main area showing argument details

5. Conclusions

The tool presented in this paper can be regarded as a proof-of-concept: it has been very useful to validate with users our idea of creating an argumentation-based system to support discussion documentation and analysis in the medical domain. The iterative and participatory design of this tool has allowed us to understand physicians' reasoning process and terminology; this way, the tool is able to foster argumentation in a natural way, being it tailored to the application domain.

The key feature that distinguishes our approach with respect to previous proposals, such as Carrell+ (Tolchinsky et al., 2006) and HERMES (Karacapilidis & Papadias, 2001), is the organization in two modules devoted to the documentation and the analysis of clinical discussions, respectively. Since well-founded structures can be exploited in the analysis phase, the first module allows the user to represent the discussion without adhering to a set of structural constraints, thus speeding up the documentation phase. On the other hand, the graphical representation allows physicians to quickly recall and continue a previous discussion. We believe that finding a good trade-off between a structured representation and a natural interaction is essential to motivate the use of the prototype, given the critical time constraints often characterizing the job of medical personnel. This has been confirmed by the evaluation with users of the part of the system devoted to the representation of medical discussions. Participants in the evaluation judged the system as very useful and coherent with their mental model of discussions; furthermore, they said that it allows deepening the reasons under decisions, thus improving comprehension of the different points of view.

The part of the system devoted to the analysis of a discussion based on argumentation schemes has not been tested with real users yet. Therefore, future work will be mainly focused on the test and revision of the features that help the user identify arguments and answer to critical questions. Attention will also be paid to summarizing the parts of the graphical representation the user is not focused in, possibly exploiting the techniques presented in (Baroni et al., 2012).

References

Aigner, W., & Miksch, S. (2006). CareVis: Integrated Visualization of Computerized Protocols and Temporal Patient Data. *Artificial Intelligence in Medicine 37(3)*, 203-218.

Baroni, P., Boella, G., Cerutti, F., Giacomin, M., van der Torre, L., & Villata, S. (2012). On Input/Output Argumentation Frameworks. *In Proc. of COMMA 2012 - Computational Models of Argument* (pp. 358-365). Vienna, Austria.

Buckingham Schum, S. (2003). The Roots of Computer Supported Argumentation Visualization. *In Visualizing argumentation*. Springer-Verlag.

Burstein, F., & Widmeyer, G. R. (2007). Decision support in an uncertain and complex world. *Decision Support Systems 43*, 1647-1649.

Compendium web site. http://compendium.open.ac.uk/index.html

Cortés, U., Vàzquez-Salceda, J., Lòpez-Navidad, A., & Caballero, F. (2004). UCTx: A Multi-Agent System to Assist a Transplant Coordination Unit. *Journal of Applied Intelligence 20(1)*, 59-70.

Fogli, D., & Guida, G. (2012). Enabling Collaboration in Emergency Management through a Knowledge-Based Decision Support System. *In A. Respicio, F.*

Burstein (Eds.), *Fusing Decision Support Systems into the Fabric of the Context* (pp. 291—302).

Fogli, D., & Guida, G. (2013). Knowledge-Centered Design of Decision Support Systems for Emergency Management. *Decision Support Systems 55*, 336-347.

French, S., & Turoff, M. (2007). Decision Support Systems. *Communications of the ACM 50(3)*, 39–40.

Glasspool, D. W, Oettinger, A., Smith-Spark, J. H., Castillo, F. D., Monaghan, V. E. L., & Fox, J. (2007). Supporting Medical Planning by Mitigating Cognitive Load. *Methods in Information in Medicine 46(6)*, 636-640.

Gordon, T. F. (2010). An Overview of the Carneades Argumentation Support System. *In Reed, C., Tindale, C. W. (Eds.), Dialectis, Dialogue and Argumentation: an Examination of Douglas Walton's Theories of Reasoning* (pp. 145-156).

Karamanou, A., Loutas, N., & Tarabanis, K. A. (2011). ArgVis: Structuring Political Deliberations Using Innovative Visualization Technologies. *In Proc. of Electronic Participation, Third IFIP WG 8.5 Int. Conf. ePart 2001* (pp. 87-98). Delft, The Netherlands.

Karacapilidis, N., & Papadias, D. (2001). Computer supported argumentation and collaborative decision making: the HERMES system. *Information systems 26*, 259-277.

Lowrance, J., Harrison, I., Rodriguez, A., Yeh, E., Boyce, T., Murdock, J., Thomere, J., & Murray, K. (2008). Template-Based Structured Argumentation. *In Okada, A., Buckingham Shum, S., Sherborne, T. (Eds.), Knowledge Cartography: Advanced Information and Knowledge Processing* (pp. 307-333). Springer.

Peleg, M., Wang, D., Fodor, A., Keren, S., & Karnieli, E. (2006). Adaptation of Practice Guidelines for Clinical Decision Support: A Case Study of Diabetic Foot Care. *In Proc. Workshop on AI Techniques in Healthcare: Evidence-Based Guidelines and Protocols*.

Plaisant, C., Mushlin, R., Snyder, A., Li, J., Heller, D., & Shneiderman, B. (1998). LifeLines: Using Visualization to Enhance Navigation and Analysis of Patient Records. *In Proc. American Medical Informatic Association Annual Fall Symposium* (pp. 76-80). Orlando, USA.

Reed, C. A., & Walton, D. (2003). Argumentation Schemes in Argument-as-Process and Argument-as-Product. *In Proc. Conference Celebrating Informal Logic @25*. Windsor, Ontario.

Reed, C. A., & Rowe, G. W. A. (2004). Araucaria: Software for Argument Analysis, Diagramming and Representation. *International Journal of AI Tools 13 (4)*, 961-980.

Rittel, H., & Webber, M. (1973). Dilemmas in a General Theory of Planning. *Policy Sciences, 4*, 155–169.

Sbarski, P., van Gelder, T., Marriott, K., Prager, D., & Bulka, A. (2008). Visualizing Argument Structure. *In G. Bebis et. al. (Eds), ISCV 2008, Part I, LNCS 5358* (pp. 129-138).

Tolchinsky, P., Cortés, U., Modgil, S., Caballero, F., & Lòpez-Navidad, A. (2006). Increasing Human-Organ Transplant Availability: Argumentation-Based Agent Deliberation. *IEEE Intelligent Systems 21(6)*, 30-37.

Toulmin, S.E. (2003). *The Uses of Argument.* Cambridge University Press.

van Eemeren, F. H., Grootendorst, R., Snoeck Henkemans, F., Blair, J. A., Johnson, R. H., Krabbe, E. C. W., Plantin, C., Walton, D. N., Willard, C. A., Woods, J., & Zarefsky, D. (1996). *Fundamentals of Argumentation Theory. A Handbook of Historical Backgrounds and Contemporary Developments.* Mahwah, NJ: Erlbaum.

Walton, R.T., Gierl, C., Yudkin, P., Mistry, H., Vessey, M.P., & Fox, J. (1997). Evaluation of computer support for prescribing (CAPSULE) using simulated cases. *British Medical Journal 315*, 791-795.

Walton, D. (2006). *Argumentation Schemes for Presumptive Reasoning.* Lawrence Erlbaum Associates.

Web Browser based Patient Care Report. Retrieved from the EMS & Fire Software! Website: http://emsfiresoftware.com/products/wpcr/

Werner, K., & Rittel, H. (1970). *Issues as Elements of Information Systems.* Working paper No. 131, Heidelberg, Germany.

Wigmore, J.H. (1913). *The Principles of Judical Proof.* Little, Brown & Co.

Wyatt, J. C., & Sullivan, F. (2005). ABC of health informatics: How decision support tools help define clinical problems. *British Medical Journal 331(7520)*, 831-833.

www.ingramcontent.com/pod-product-compliance
Lightning Source LLC
Chambersburg PA
CBHW051045160426
43193CB00010B/1073